Cornell Publications in the History of Science

THE
HIPPOCRATIC
TRADITION

CORNELL PUBLICATIONS IN THE HISTORY OF SCIENCE

The Hippocratic Tradition
WESLEY D. SMITH

The Correspondence of Marcello Malpighi. 5 volumes.
HOWARD B. ADELMANN

The Baglivi Correspondence from the Library of Sir William Osler
DOROTHY M. SCHULLIAN

— *Galenism: Rise and Decline of a Medical Philosophy*
OWSEI TEMKIN

Galen on the Usefulness of the Parts of the Body. Translated from the Greek with an introduction and commentary. 2 volumes.
MARGARET TALLMADGE MAY

Marcello Malpighi and the Evolution of Embryology. 5 volumes.
HOWARD B. ADELMANN

The Embryological Treatises of Hieronymus Fabricius of Aquapendente. A facsimile edition, with an introduction, a translation, and a commentary. 2 volumes.
HOWARD B. ADELMANN

THE
HIPPOCRATIC
TRADITION

Wesley D. Smith

CORNELL UNIVERSITY PRESS
Ithaca and London

First published 1979 by Cornell University Press
Published in the United Kingdom by Cornell University Press Ltd.,
2-4 Brook Street, London W1Y 1AA.

International Standard Book Number 0-8014-1209-9
Library of Congress Catalog Card Number 78-20977
Printed in the United States of America
*Librarians: Library of Congress cataloging information
appears on the last page of the book.*

CONTENTS

PREFACE

This book is concerned with thought about medicine, not with medical practice, and I have not addressed the subject of medical ethics as related to Hippocrates and the Oath. Particularly, I am interested in the creation of ways of thinking and talking about history, in modern and in ancient times. Traditions of interpretation are often most clearly recognizable for what they are when they are demonstrably wrong. Hence, there is considerable emphasis on aberrations that the Hippocratic tradition has produced. When the shape and substance of the tradition become clear, one can then accept what in it is worth accepting and learn much from the errors one is trying to overcome. I intend this book to serve as a prolegomenon to a new reading of the Hippocratic collection.

Hippocrates has been a hero to the people of several cultures. Consequently, his interpreters and admirers have emulated, or in some cases one is tempted to say imagined, in him qualities that reflected their own ideals. This present book comes out of my own need to make sense of the cacophony in the subject and to determine why Asclepiades, Rufus, Galen, Boerhaave, and Sprengel responded to him as they did, and whence came their passions on the subject. In the process I have tried to seek out and assess all relevant evidence relating to interpretations of Hippocrates. The several new theses I advance will, I hope, seem compelling in the perspective of the whole tradition.

I have chosen to begin with modern views of the Hippocratic tradition (1500 to the present), to deal with their bases in Galen's

7

interpretation of the ancient tradition (material written ca. 150 to 200 A.D.), and then to reconsider that ancient tradition (ca. 400 B.C. to 200 A.D.). My discussion of modern views assumes that the reader has some preliminary understanding of the material I will deal with, explain, and even refute in my subsequent chapters. Reversing the order would not have solved the problems. I have had to rely on the reader to pursue what needs clarification, either in this book through the use of the index or elsewhere in the works suggested in the bibliography. I have hoped that the audience would be a broad one, and hence have not discussed technical material that was not immediately relevant.

My argument is cumulative. The treatment of post-Renaissance attitudes, in the first chapter, is intended to show how constrictive intellectually this tradition has been and how readily scholarship can become the factory of evidence for the current faddish view.

The second chapter, on Galen's Hippocratism, examines similar phenomena at an earlier stage in the tradition and develops the information necessary for rewriting the earlier history of medical thought in its relation to the Hippocratic Corpus. In order to pursue Galen's notions of the history of medicine through his numerous and complex writings I have ventured an intellectual biography of him focused on his relation to Hippocrates. Information he offers about his life and about the composition of his works can shed much light on his presentation of matters of historical fact and can resolve many of his notorious contradictions. Furthermore, precise and thorough assemblage of Galen's historical statements provides the basis for a number of novelties in the reconstruction of the early history of medical thought.

In the third chapter, I have attempted a continuous synthetic account of thought about the history of medicine and about the Hippocratic Corpus in the pre-Galenic period. My corrections of traditional versions of that history follow from the revelations in the first two chapters.

It will be apparent how indebted I am to the people with whom I disagree: Deichgräber, Diller, Bardong, Ilberg,

Wellmann, Jaeger, Edelstein, Lonie, and many others, including, of course, Galen, the fount of false as well as true medical history. I wish to acknowledge here also debts to people with whom I have not ventured to disagree: Volker Langholf for many corrections of the manuscript; Owsei Temkin for his benign encouragement; Lloyd Stevenson, Janet Koudelka, and others at the Institute of the History of Medicine in Baltimore; Phillip DeLacy, colleague, friend, and exemplar; and Bernard Knox and the Institute for Hellenic Studies, where I conceived this book after I had set out to write another one. I also thank the John Simon Guggenheim Foundation for support of my leisure for writing, and my wife, Karen, for help such as only she could give.

WESLEY D. SMITH

Philadelphia, Pennsylvania

THE
HIPPOCRATIC
TRADITION

1

THE MODERN
HIPPOCRATIC TRADITION

When the Renaissance brought renewal of direct knowledge of the ancient Greek medical texts, Hippocrates became known both through the writings of the Hippocratic Corpus and through the works of Galen. There were no obvious reasons to disbelieve Galen's claims that he followed the teachings of Hippocrates accurately and that he understood the works of the Corpus and knew which were "most genuine" and which were spurious and unlike the outlook of Hippocrates. Renaissance scholarly work took Galen's views of Hippocrates and Hippocratic medicine as its guide, and took for its object the better understanding and emulation of the Divine Galen as well as the Divine Hippocrates. Since the early Renaissance had seen a considerable improvement in scientific medicine as a result of the importation of Greek learning as translated and refracted through Arabic sources, one of the concerns of Galenists and Hippocratists in the later Renaissance was to use the original Greek sources to correct and purify Greek medicine in its Arabian dress.

However when, little more than two centuries afterward, Emile Littré produced his fine edition and translation of the Hippocratic Corpus which has been the basis for virtually all modern work, entirely different assumptions are obvious. Galen's reputation as scientific medical man had sunk drastically and with it Galen's credibility as interpreter and guide to scholars in Hippocratic studies. Littré considered the spirit of Hippocrates quite antithetical to that of Galen, and the works on which his interpretation of Hippocrates rests are different from those on which Galen's opinions were based.

The change in opinion about Hippocrates and the Hippocratic writings did not occur as the result of new information, nor is it clear that people were aware, or are now aware, how extensive it had been. The reversal in point of view is almost completely obscured in Littré's discussion of the preceding scholarly tradition, where he assumes that the point of view he brings to criticizing his predecessors is the natural one. Indeed, it was natural in Littré's time and still is insofar as we are the intellectual children of the Enlightenment. I shall try to determine by what process the change of view occurred and to estimate its significance for our understanding of the ancient texts and ancient medical tradition. I will conclude that the medical men led the way, generating medical "history" out of their own current scientific interests instead of out of historical study and that subsequently philologists and historians, infected by the progressive attitude and euphoria of the scientists, set out to find the evidence with which to sustain the medical men's views.

FROM THE MEDICAL MEN TO
THE HISTORIANS

The crucial first step in the process was dissociation of Hippocrates from Galen in "spirit" and doctrine. In that, it appears, the alchemist and mystic Paracelsus (1493–1541) led the way when he claimed that Hippocrates had had the true medical spirit but that Galen and other sophists had perverted his views. Paracelsus' specific positive views were not long influential, but his attack on old authority and structures of thought stimulated others.[1]

Paracelsus, as Walter Pagel put it, "broke away from the ordi-

1. Owsei Temkin, *Galenism* (Ithaca, N.Y., 1973), pp. 129–130, writes, "Though it would be wrong to say that the separation of Hippocrates from Galen was the achievement of Paracelsus, he certainly contributed to it, and thus the main responsibility for traditional medical science was assigned to Galen."

nary logical and scientific ratiocination, ancient and medieval and modern,"[2] to pursue a direct understanding of the nature of things: believing that man the microcosm can, through his correspondences with the macrocosm, know the truth. Whether or not he burned Galen and Avicenna publicly to initiate the course of medicine he taught at Basel, Paracelsus continuously attacked the deadening influence of authority. His predecessors, the humanists who had revived ancient learning, had come to scorn the scholastic mind as slave to the authority of books, never looking at life. But the humanists had gone directly to the fountain to drink, by making the original Greek writings available and by imitating their spirit. Paracelsus carried that process one step farther: for him all ancient authorities, especially Aristotle and Galen, were as guilty as the medieval scholastics, guilty of turning their backs on nature and in the process losing all effectiveness.

But Hippocrates fared much better: he was the last true physician before Paracelsus himself. As Paracelsus saw the matter, God had allowed medicine to become known through Apollo, Machaon, Podalirius, and Hippocrates; in them the light of Nature shone forth. But the Evil One interfered and caused medicine to fall into the hands of the antiphysicians so that it became entangled with persons and sophistries.[3] Paracelsus argues that Hippocrates (and his mythical predecessors) proved himself through his works. Although Paracelsus is never more specific, he seems to refer to the curing of plagues and healing of kings at will, that is to say the mythology contained in the *Letters, Speech from the Altar,* and other pseudepigrapha. The original purity of Hippocrates' outlook was overwhelmed and obscured by the legion of chattering sophists who followed. Paracelsus will not permit himself to let the fact that the sophists

2. Walter Pagel, *Paracelsus* (Boston and New York, 1958), p. 50.
3. *Seven Defensiones,* Preface, in *Four Treatises of Theophrastus von Hohenheim, Called Paracelsus,* p. 10, in *Theophrastus Paracelsus Werke,* ed. Will E. Peuckert (Basel, 1965–1968), II, 497. Machaon and Podalirius are the mythical sons of Asclepius, the original Asclepiads who, according to the *Iliad,* accompanied the Greek army to Troy.

are so numerous deter him! "From it arises the error that Hippocrates must be a gossip and the spirit of medicine must become a sophistic chatterer."

By rejecting all interpretations by the followers of Hippocrates, Paracelsus appears to be promising new insights into the Hippocratic writings. But in accord with his attack on authoritative books and scholasticism, Paracelsus did not engage in bookish dispute with the previous interpreters, nor, as far as I can determine, did he study the Hippocratic text to offer his own interpretations in support of his assertion that Hippocrates healed "with the true spirit of medicine."[4] He depended on "what everyone knows," I suspect. Nevertheless, perhaps as a relic of his brief tenure as a teacher at Basel, he left the beginnings of a commentary on the *Aphorisms,* in which he finds reflections of his own views in Hippocrates' simple observations and advice.[5] Commenting on the first phrases of the first aphorism, "Life is short, the art long," he speaks at length of the knowledge of all nature and the cosmos which the true physician must have, as Hippocrates knew. The following phrase in the Latin on which he comments is *tempus acutum.* Paracelsus takes it to mean "weather is dangerous." He comments, "Wherefore the physician must be an accomplished astronomer" (*darumb sol der arzt ein erfarner astronomus sein*). The actual meaning of the Greek, *kairos oxys,* is "the chance is on the razor's edge," or "opportunity is fleeting," or the like. In the text Paracelsus is interpreting, *tempus* was used to translate *kairos* as well as to translate *hôrên* in *Aphorisms* 1.2, a word that does mean "season" or

---

4. *Seven Defensiones,* Conclusion, in *Four Treatises,* p. 41. The accomplishments of Apollo, Machaon, and Podalirius are healing with the true spirit of medicine, performing *Prodigia, Signa,* and *Opera.*
5. The commentary is in German, in *Sämtliche Werke,* ed. Karl Sudhoff (Berlin, 1931), 1 Abteilung, 4 Band, pp. 493–546. Paracelsus comments on *Aph.* 1.1–25, and 2.1–6. I have found only two discussions of this commentary: Walter Brunn, "Beträchtungen über Hohenheims Kommentare zu den Aphorismen des Hippokrates," *Nova Acta Parcelsica* 3 (1946), 24–42; and L. Braun, "Paracelse, Commentateur des Aphorismes d'Hippocrate," in *La collection hippocratique et son rôle dans l'histoire de la médecine* (Leiden, 1975), pp. 335–346. Both point to profound differences in point of view between commentator and text under consideration.

"weather." As a rendering of the Latin, without reference to the Greek original or to commentaries, Paracelsus' interpretation does make sense, though it is wrong because it lacks contact with the original text and its concepts.

For the most part, Paracelsus' comments are paraphrases and expansions in his own terms of what *Aphorisms* says about good and bad symptoms, the necessity for fitting treatment to circumstances, and the like. He does not attempt to draw Hippocrates' "philosophy" out of the *Aphorisms,* nor does he refer to other works of the Corpus or to other interpreters. In some few places the terms in which he interprets Hippocrates recall his earlier generalizations about the light of Nature that shines through him. For example, on *Aphorisms* 1.11, where the text, as translated by W. H. S. Jones, says, "lower diet during exacerbations, for to give food is harmful," Paracelsus comments that the meaning of the aphorism is that if Nature is unwilling to take food, food should not be given.

As modern reinterpretation and revaluation of Hippocrates, these are very small beginnings, but Paracelsus' generalizations about Hippocrates' spirit stimulated others, whose reinterpretations became more specific. J. B. van Helmont, (1577–1644), who developed Paracelsus' iatrochemical notions and systematically attacked Galen's humoralism, repeated Paracelsus' praise of Hippocrates and found specific support for it in *Ancient Medicine:* that work, more correctly than Galen, had said that "diseases are not hot or cold, but something acid, sharp, bitter and biting."[6] Hippocrates was not to have much of a career as an iatrochemist, but the treatise *Ancient Medicine,* for its methodological discussion, was to have a large place in later reinterpretations of Hippocrates. Van Helmont read the Hippocratic letters which describe the successful cure of a plague, and he remarked that, despite their increasing sophistication, Galen and all later physicians had lost that early insight. "Hippocrates had less garrulity but more candor, science, and heavenly light" (*Hippocrates*

6. *Blas Humanum* 52. Johannes Baptistus van Helmont, *Ortus Medicinae* (Amsterdam, 1648), pp. 190–191.

minus garrulitatis, plus candoris autem, scientiae caelestis luminis habebat).[7]

From about the time of Paracelsus the authority of what had been the Hippocratic-Galenic system of medicine dwindled rapidly. Between Andreas Vesalius' *Fabrica* (1543), which showed that Galen was not final in his anatomical method and observations, and William Harvey's work on the circulation of the blood in 1628, which showed Galen's physiology to be wrong, attacks ranged from criticism of the comprehensiveness and consistency of the ancient system, as by Jean Argentier (1513-1572), to the general condemnation by Francis Bacon (1561-1626), who was prepared to do away with ancient systems because they were "fruitful of controversies but barren of works."[8]

Bacon distinguished Hippocrates from all others in his survey of the condition of medical science and lamented "the discontinuance of the ancient and serious diligence of Hippocrates, which used to set down a narrative of the special cases of his patients and how they proceeded, and how they were judged by recovery or death."[9] Bacon praised Hippocrates' methodology as appropriate to the new science he envisioned and thus foreshadowed the direction the new Hippocratism was to take. Bacon may have been influenced by the work of Guillaume de Baillou (1538-1616), a professor in Paris, who had done research of the kind that Bacon thought desirable. Not a critic of the Galenic system, Baillou was one of the "conciliators," who explained away apparent contradictions within the Galenic system and between Galen and Hippocrates. In his gynecology and his *opuscula medica* he draws widely from the Hippocratic Corpus

7. *Tumulus Pestis* ch. 2, *Ortus Medicinae*, p. 14. In ch. 18 of the same work he implies that Hippocrates' remedy was a compound of sulphur, a method he hoped to rediscover (*Ortus Medicinae* p. 67).
8. For Argentier's criticism of Galen, which was independent of that of Paracelsus, see Pagel, *Paracelsus*, pp. 301-304, with references, and Temkin, *Galenism*, pp. 141-142. Francis Bacon, *Great Instauration*, Proemium, in the *Works of Francis Bacon*, a new edition with a Life of the Author by Basil Montagu (Philadelphia, 1843), III, 334.
9. "Of the Advancement of Learning," 2.10.4, in *The Advancement of Learning and New Atlantis*, ed. Arthur Johnston (Oxford, 1974), p. 108.

and from Galen's works, which he finds lucid and correct, and he defends their doctrine against criticisms by his predecessor Jean Fernel.[10] Baillou was influential, however, because of his original research in the manner of Hippocrates' *Epidemics*, which he published in *Epidemiarum et Ephemeridum libri duo.*[11] These were "constitutions" in the Hippocratic manner—reports of the year's weather with accounts of the diseases that were typical, along with comparison with the observations of Hippocrates himself. Paracelsus had argued that diseases were related to place and time and that a German could not depend on observations and inferences made in the Mediterranean long before. Baillou was offering an example of new research in nosography, which could be the basis of a new or reformed empirically based science.

Bacon's ideas of scientific methodology influenced Thomas Sydenham (1624–1689), the "English Hippocrates," who was associated with John Locke and with Robert Boyle and other founding members of the Royal Society.[12] Sydenham pursued the notion of a "natural history of disease" to be arrived at by meticulous recording of observations. He considered that his function as physician was to cure, not to develop theories or to find causes, and in treatment he was expectative rather than active. He was opposed to the traditional course of medicine which Oxford had offered him. Circumstances of the Oxford he attended, in the wake of the first civil war, in which he had served in the Parliamentary cavalry, were favorable to his outlook: puritan, antischolastic, antiroyalist. He became a bachelor of medicine without having to pursue the lectures on Galen, the public dissections, or the formal disputations. Kenneth Dew-

10. *De virginum et mulierum morbis liber, in quo multa ad mentem Hippocratis explicantur quae et ad cognoscendum et ad medendum pertinebunt,* and *Opuscula Medica . . . in quibus omnibus Galeni et veterum authoritas contra J. Fernelium defenditur.* I have seen these works only in the Paris edition of 1643, edited by M. Jacobus Thevart.
11. I have seen this work only in the Paris edition of 1640, edited by M. Jacobus Thevart.
12. Kenneth Dewhurst, *Dr. Thomas Sydenham, His Life and Original Writings* (Berkeley, 1966), pp. 62–63. I am indebted to Dewhurst's book for my initial acquaintance with much of my information about Sydenham.

hurst quotes from the diary of a contemporary student a description of Sydenham's practical attitude: "Physick says Sydenham, is not to bee learned by going to Universities, but hee is for taking apprentices; and says one had as good send a man to Oxford to learn shoemaking as practising physick."[13] Yet Sydenham admired and emulated Hippocrates—essentially Bacon's Hippocrates, the careful observer—and by his own success contributed to an emerging Hippocratism opposed to the traditional Galenism. For the new Hippocratism at its origins, then, "Hippocrates" was a spirit and a method rather than a body of doctrine. Scholarship on the Corpus and rewriting the history of early medicine to justify such use of Hippocrates were yet to come.

Medical history was in its infancy. The *Histoire de la médecine* of Daniel Le Clerc, written just before 1700, is particularly interesting for its definition of medical history and also for its reflection of old and new attitudes about early medicine in this transitional period.[14] Le Clerc is sensitive to new ideas in medicine, concerned for the honor and dignity of the profession as well as for its progress, and hopeful that the best aspects of the ancients not be lost in the pursuit of novelty, but also that the ancients not be treated as "oracles" and that medical science be based on the proper combination of observation and reasoning. He says that his purpose as historian is to enter sympathetically into each period and to report objectively, not simply to serve his own taste (Preface). With the assistance of previous scholarship, whose prejudices he approaches in a critical spirit, he claims to write the first comprehensive history of medicine, as opposed to a series of biographies of physicians. He intends a sequential narrative of the development of the science, emphasizing "les principaux *raisonnements,* et les *experiences* les plus considerables" (the most important theories and observations). He is especially con-

13. Ibid., p. 17.
14. Le Clerc's first edition, published in Geneva in 1696, carried the history of medicine through the time of Galen. My page references in the text are to the expanded edition (The Hague, 1729), which adds a sketch for continuation from Galen to the mid-sixteenth century, pp. 765–820. The pages of the Preface, in which he describes his approach, are unnumbered.

cerned to interpret Hippocrates from his own writings, not from what others, in their enthusiasm, have attributed to him (Preface). His considerable ambition, high ideals, and great labor did not prevent him from reproducing the commonplaces and contradictions characteristic of his period. His summary statement about Hippocrates is that "la Médecine d'Hippocrate roule toute sur l'Observation. Ce Chef des Médecins s'est plus attaché a faire des expériences, qu'à pousser fort loin son raisonnement, quoi qu'il soit l'un des premiers qui ont rendu la médecine raisonnée" (Hippocratic medicine entirely revolves around observation. The best of physicians is more devoted to doing experiments than to extending his theory, though he is among the first to have made medicine rational. Preface, see also p. 705). Le Clerc thus offers the evaluative generalization which was to determine the course of Hippocratic scholarship. But his evaluations are superficial. The Hippocrates he finds in reading the Corpus is still very much the Hippocrates of Galen. Hippocrates' philosophy, as interpreted by Le Clerc, is based on the notion of just Nature (which virtually equals heat), which governs the organism as it governs the macrocosm and uses the natural faculties of attraction, retention, preparation, and expulsion to administer the body's economy (pp. 115–117). He outlines Hippocrates' view of the sources of health and disease according to the four-humor, four-quality theory of the *Nature of Man,* but he adds that *Ancient Medicine* seems to talk of an infinity of humors (pp. 144–145, quoting the passage from chapter 15 which van Helmont had used). Hippocratic therapy, based on allopathy, comes out of Hippocrates' view of Nature: one might think from the *Epidemics* that Hippocrates was a spectator of Nature's efforts, without much interference; in fact, he engaged in active therapy, but still to a lesser extent than physicians in subsequent ages did, says Le Clerc, and he proceeds to draw from the therapeutic works of the Corpus indications for use of diet, purging, bleeding, and drugs to alter the body's constitution (pp. 191–232). As to bleeding, Le Clerc argues at length that Galen is wrong about Hippocrates' habitual and extensive use of bleeding in fevers and that Johannes Riolanus, the famous Galenist

(1580–1647), erred in his interpretation of Hippocrates' venesection as well as in his argument that there is indication of knowledge of the circulation of the blood in Hippocrates (pp. 202–208, 128 with note 6).

For details of the "Hippocratic Question," Le Clerc refers his reader to Hieronymus Mercurialis, Anutius Foës, and others who had dealt with it.[15] But he considers it an established fact that the contradictions within the Corpus are accounted for by multiple authorship and that the most theoretical of the treatises are the most suspect (p. 240). In his comparison of Galen's medicine with that of Hippocrates, he agrees with Galen's claims to have followed Hippocrates' principles in virtually all respects, but he argues that the different emphases, Hippocrates' on *expérience,* Galen's on *raisonnement,* are of crucial importance. Further, he adduces the treatise *Ancient Medicine* as evidence that the rigidity of the four-humor system as Galen uses it is not characteristic of Hippocrates (p. 715). Le Clerc sees no reason to suspect the genuineness of *Ancient Medicine,* regardless of Galen's views on it. Great as he was, Galen demonstrably read into Hippocrates some things that were not there.[16]

Le Clerc's interpretations show a large degree of inconsistency, especially from the retrospect of later discussions. They might be considered as simply a bad compromise forced on him by the contradictions among his authorities. At a later time it would be obvious to anyone that *Nature of Man* and *Ancient Medicine* cannot be reconciled (even by assuming a flexibility of view in *Nature of Man*), and that the Natural Faculties, if they are inferred from *Nutriment,* cannot be found in the other works. So

15. Hieronymus Mercurialis, *Censura operum Hippocratis* (Venice, 1585). This edition differs somewhat from the 1588 edition as reported by Littré. Mercurialis attempts to make consistent judgments on genuineness of the Corpus following Galen's leads. Foës, in his edition of 1595, rejected as spurious the letters and other nonmedical addenda to the Corpus, but was generally restrained in offering judgments about the genuineness of the medical works. He was, however, suspicious that there were Cnidian works in the Corpus. On this see Iain M. Lonie, "Cos versus Cnidus and the Historians, Part 1," *History of Science* 16 (1978), 48–50.
16. Mercurialis put *Ancient Medicine* in his fourth class, works that are alien to Hippocrates and his school.

we might reason. But when I was considering Le Clerc's views, and wondering how much inconsistency might be tolerable, I was reassured by thoughts of Sydenham. He claimed to work without any theory at all and found it sufficient in his exposition of his new, cooling regimen for smallpox to say that it did cure, while the old theoretically based treatment did not; yet when writing of gout he dealt with it in terms of the four constitutions, and he adopted a corpuscular theory of epidemics.[17] In the period of Le Clerc and Sydenham, when the notions of new empirical scientific method were just being explored and made consistent, a man like Le Clerc could be guided in his vision of what Hippocrates may have believed only by his own sensibilities; he would therefore produce what earlier or later would look like a muddle. In subsequent discussions of Hippocratic medicine, we can observe the sorting and resorting of the elements in Le Clerc's discussion until consistency is achieved. Each one who takes up the subject seems to have an empathy for Hippocrates which is turned into a sense of insight into his medicine.

Hermann Boerhaave (1668–1738), who made such great contributions to modern medicine and medical education, appears to me to have made a decisive contribution also to views of medical history and to Hippocrates' place in it. He was an eloquent preacher of Hippocratism and modernism. His books and students were influential all over Europe. His own temperament shines through his historical constructions. In his education Boerhaave had pointed himself in two directions, toward medicine and toward ministry in the Dutch Reformed church. According to his sketch for his biography (in which he speaks of himself in the third person), he first read the Bible for himself and then read chronologically through the fathers of the church. He found the simple and pure doctrine of the early fathers admirable, but he

17. See especially Dewhurst, *Sydenham,* pp. 66–67, 140–144, for Sydenham's various inconsistencies. For his practical defense of his cooling regimen and his criticism of the old theory, see "Smallpox, 1669," pp. 101 ff., especially 104–106.

regretted that the subtleties of the Schools had later corrupted theology. He greatly regretted that the interpretation of the Holy Scripture was sought for among the sects of the Sophists; and that the metaphysical reflections of Plato, of Aristotle, of Thomas Aquinas, of Scotus, and—in his own time—of Descartes, were considered as laws according to which the views of God expressed in the holy scriptures should be amended.[18]

Similarly, he describes his education in medicine:

> It is perhaps incredible that he did not attend lectures by any professor in medicine, except for a few by the celebrated Drélincourt shortly before his death ... He began his reading of the ancient medical writers in chronological order, starting with Hippocrates; soon he understood that the later authors owed to Hippocrates everything that was good in their work; therefore to him alone he devoted a long time, reading him, summarizing and analyzing him. Running through the more recent writers, he halted at Sydenham, whom he worked through several times, each more eagerly.[19]

The style of medical education which Boerhaave fostered at Leiden, which he hoped would continue after his death, he called a Hippocratic School.[20] His inaugural lecture in 1701 was entitled *de commendando studio hippocratico*. While he called himself a Hippocratic, Boerhaave was up to date in virtually everything. He pressed for a unified medical science empirically and experimentally based. His own reform of medical education away from study of authors toward study of fields, chemistry, botany, etc., was a move toward unified science.[21] For models he commended Vesalius, Fallopius, Harvey, Bacon, Newton, Boyle, and in their company Hippocrates and the Hippocratic clinician Sydenham. At the opening of his *Institutiones Medicinae,* he gave

18. From his autobiographical notes, translated in G. A. Lindeboom, *Herman Boerhaave, The Man and His Work* (London, 1968), p. 380.
19. Ibid., p. 381.
20. Ibid., p. 386.
21. On this reform, see Temkin, *Galenism*, pp. 173-174; and Lindeboom, *Herman Boerhaave*, pp. 361-374.

a sketch of the history of medicine. The *Institutiones* was a brief outline of his medical lectures intended for purchase by the students. First in pirated, then in authorized editions, it was widely sold and used as a textbook in Europe and in the New World in the eighteenth century.[22]

He begins his history of medicine with the body's natural tendency to heal itself and the mind's natural pursuit for remedies against disease and discomfort (sections 1–5).[23] In time, he says, the piling up of experience was followed by increasing precision in recording that experience: descriptions of diseases and of materia medica and their effects (secs. 6–9). At the same time, knowledge of the body increased from the inspection of entrails, mummification, treatment of wounds, and the like (sec. 11). The only hindrance to the natural progress of medicine that Boerhaave envisions in the early stages is the restriction of medical practice to particular families and to priests (10). Finally, says Boerhaave, Hippocrates unified all aspects of the Art, created a unified corpus of Greek medicine, and first earned the name of genuine physician. Combining experience (*empeiria*), reasoning from experience (*analogia*), and chaste wisdom (*casta sophia*) he founded dogmatic medicine for all the ages (13). The medicine founded by Hippocrates was cultivated by the Asclepiads, was made orderly by Aretaeus of Cappadocia, and was made more precise by various physicians, especially those of Alexandria, until it came to Galen (14). Galen collected all the scattered material and ordered it, explaining it according to peripatetic doctrines. By his efforts he did much good for medicine, but he did at least as much harm because he was the source from which everything in medicine was explained by elements, by the qualities commonly called cardinal, by their gradations, and by the four humors—all this with much more subtlety than truth (15).

From this estimate of Galen, Boerhaave passes to the extinc-

22. See G. A. Lindeboom, *Bibliographia Boerhaaviana* (Leiden, 1959), pp. 27–40. For the *Institutiones Medicinae* I have used especially the 1713 authorized Leiden edition. There are few and insignificant changes of the prolegomena in the other editions that I have inspected.

23. References are to the sections of the prolegomena of the *Institutiones* as numbered by Booerhaave.

tion of learning after the sixth century. To the Arabs between the ninth and thirteenth centuries he attributes contributions to surgery and materia medica, but says that they infected the art even more than before with the faults of Galen (16). At length two means of reform and correction appeared: the revival of Greek and Greek medical writings, which resulted in the reestablishment of Hippocratic teaching in France, and the beginnings of chemical and anatomical experiments (17).[24] Finally, the immortal Harvey overturned all the theory of his predecessors by his demonstrations and laid down a new and more certain basis for medicine (18). Boerhaave summarizes: the most ancient art consists of collection of accurate observation. Subsequently, there was reasoning about the causes behind experience through rational disputation. The first is stable and dependable, the second changing and doubtful, dependent on one's sect. Proper procedure can create what is needed: a science which is rational as well as empirical (19).

Boerhaave does not have new theories about the Hippocratic Corpus and its preservation. Indeed, it is difficult to distinguish his opinions about it from Le Clerc's. Unlike Le Clerc, however, he offers an incisive valuation which is at once practical and idealistic, not a gesture of fairness toward all views, but a ruthless assertion of where the true method lies, and who are the anti-progressives (more subtle than Paracelsus' devil-inspired antiphysicians, though similarly manichean in tendency). Boerhaave's essential history of medical science sounds modern: that is because he offered the themes that medical historians would develop. As Charles Lichtenthaeler has observed, the historians of medicine came to imitate the scientists in their notion of true method, their optimism, and their emphasis on the basic importance of the things that have survived or been influential.[25] It is not surprising that Boerhaave, who had such an effect on ideas about medicine, should have affected medical history.

24. Boerhaave does not specify to whom in France he refers, whether Guillaume de Baillou or others.
25. Charles Lichtenthaeler, "De quelques changements dans notre conception de l'histoire de la médecine," in *Deux Conférences* (Geneva, 1959), see especially pp. 38–45.

What is remarkable is the ease with which his outlook became *the* view, as though there was no other way in which the past could be conceived.

The view of early medical history around 1800 is well exemplified by the *Versuch einer pragmatischen Geschichte der Arzneikunde* in five volumes by the learned Kurt Sprengel, who, early in his career, wrote an introduction to Hippocrates and the Hippocratic Question.[26] For Sprengel, the methodological and technical revolution effected by Hippocrates is much clearer, but Hippocrates' relation to the Hippocratic Corpus is much vaguer than it was for Boerhaave. It is a sign of the time when he wrote that Sprengel compares the problem of the Hippocratic writings to that of the Homeric poems (I, 376): F. A. Wolf's *Prolegomena ad Homerum* was published in 1796. The fundamental contribution of the historical Hippocrates, as Sprengel reconstructs him, was to show that the methodology followed by the Asclepiads and philosophers before his time was incapable of perfecting the science. "He showed physicians that their first duty was to observe carefully the progress of nature. He showed the uselessness of theories and proved that observation alone is the basis of medicine" (I, 427). If men had followed the route Hippocrates laid out, "Greek medicine would have attained in a brief time a degree of perfection which we can hardly imagine, since anatomy, which was not slow to develop, would have shed more vivid light on it. But the bright hopes were not realized. Simple observation was repugnant to the dominant spirit of the century, and anatomy served only to confirm the speculations and theories of the dogmatic physicians" (I, 349). Sprengel clearly believes that these dogmatic physicians included Hippocrates' own sons and son-in-law, who falsified and interpolated his work so that we can hardly discover what is his. Hippocrates' doctrine can only be recovered by subtracting from the Corpus all

26. Sprengel's early work on Hippocrates is called *Apologie des Hippokrates und seiner Grundsätze* (Leipzig, 1789), a translation and discussion of several works. For his *Versuch einer pragmatischen Geschichte der Arzneikunde* (Halle, 1792–1799), page references in the text are to the more readily available second edition, five volumes (Halle, 1800). I have not seen the 1792–1799 edition.

Platonism, Aristotelianism, and other subtle philosophizing (I, 382–384). Finally, Sprengel thinks that the humoral and elemental theory of the first part of *Nature of Man* must express Hippocratic views. Indeed, Hippocrates did not write it. Galen is wrong to say that Plato cites that work; in fact, the work which Plato cites must be lost, but the philosophy of microcosm-macrocosm to which Plato refers must be similar to that expressed in *Nature of Man* (How often similar reconstructions were to be offered as original ones in the following two centuries!). Hippocrates' original contribution was in semeiotics, dietetics, and, a factor which Stoll and Lepecq, among others, appreciated, in charting the courses of diseases through their critical days (I, 384–390).

Before considering various specific nineteenth-century developments in Hippocratic scholarship and medical history, let us skip forward to the highly respected medical history by Max Neuburger, which was written about a century after that of Sprengel,[27] to observe the further evolution of the history of medicine we are tracing. By 1906, Hippocrates' unique scientific insight is very clear, but his works have disappeared entirely. Neuburger's construction starts from the Greek debt to older cultures. From Egypt and Mesopotamia the Greeks inherited a knowledge of drugs together with many fundamental ideas of wide theoretical significance. Because of keen competition between different cultural centers the Greeks managed to develop their heritage. They were free from priestly domination. Hence everywhere in Greece medicine developed beyond formal dogma and mere empiricism (p. 188). The different schools took their own paths: the Cnidians worked on the descriptions of the numerous types of disease; the Coans sought to unite all disease into a conceptual unity subject to prognosis; the Sicilians attempted to arrive at first principles by way of natural philosophy, whence they could deduce the theory of disease and the basis of medical treatment (pp. 113–120). In the time of Hippoc-

27. Max Neuburger, *Geschichte der Medizin* (Stuttgart, 1906). The first German edition stops with the time of Galen. Page references in the text are to the English translation by Earnest Playfair, *History of Medicine* (London, 1910).

rates, Greek medicine arrived at a critical period which called for a leading spirit: fanatical empirics on the one side and subjective iatrosophists spinning hypotheses on the other made it seem inevitable that medicine should lose itself in a sea of speculation or be overwhelmed in a waste of barren formalism (p. 128). But Hippocrates appeared. Unique and heroic, he freed himself from oriental dogmatism and from priestly caste and "was capable of climbing alone the steps of rational science and moral dignity" (p. 129). "To Hippocrates, and to him alone, was it given to use freedom with discretion: to recognize and with wise restraint make no attempts to overstep the limits of human knowledge; to eliminate alike the dogma of caste and the element of uncontrolled speculation" (p. 188). As Neuburger views the matter, we cannot know "to which pathological system Hippocrates inclined, nor is it of the slightest importance since his medical activity was only to the slightest extent influenced thereby" (p. 139). It is the "conception of medical vocation and the method of medical thought and action, true now as then," which raises Hippocrates' creed to "the highest pinnacle of Greek medicine and even makes it the well-spring of medical science for all time" (p. 124).

But Hippocrates' uniqueness was unfortunate: no one, not even his sons and students, understood him, as we can see from the Corpus itself.

> Unfaithful to the dispassionate, purely clinical intellectual method of the great Coan, a considerable portion of the Hippocratic collection is pervaded by the speculative spirit, proof that his disciples and their pupils aimed either at bringing the practical principles of Hippocratism into harmony with *a priori* ideas derived from natural philosophy or at reinforcing Coan fundamentals with the physiological and pathological theorems of other schools. Striving to outdo the master, to dress his empirical dicta in a garb of pseudo-science, the tendency grew to consider as vital what to him was unessential and accessory, and thereby only too frequently to lose sight of the gist of his teaching." [P. 160]

The work *Nature of Man*, which was written by his son-in-law, shows how soon the betrayal occurred.

Once Neuburger has established his principles in this way, he has turned the Corpus into a Smorgasbord from which he can select the elements which appeal to him in each treatise without concern for the structure of the whole: virtually no work of the Corpus is wholly uninfluenced by Hippocrates, but all or virtually all is corrupted by his successors (p. 123). Hence Neuburger picks out the Hippocratic sentiments in *Aphorisms, Ancient Medicine, Prognostic,* and also in what he considers the later, more corrupted works: all have bits appropriate to the "echt nüchterne Sinn ihres intellectuellen Urhebers" (the genuinely straightforward view of their intellectual source), and all have examples of rashness, failure of observation, and speculative quality. "The Hippocratic ideal, however seldom realized, yet lives, unfettered by doctrines, throughout all time" (p. 160).

Some of the heightened drama in the story Neuburger tells appears to be his own addition; in the main, however, his version of what is Hippocratic and how to find it differs little from other standard medical histories: those by Arturo Castiglioni, Max Wellmann, Charles Singer, and others. What I wish to point out here is that this story about medicine in the time of Hippocrates is in fact an etiological myth, an analytical scheme dressed up as a narrative of events. The analytical scheme is intended to explain the inspirational value of the Hippocratic writings, along with the inconsistency and inarticulateness of their "scientific spirit." The form of the myth comes out of the Enlightenment, for which the enemies of progress were clearly identifiable as priests, tyrants, and philosophical speculation, whereas progress comes from the unique genius, who puts his mind to Nature as it is commonly observed by all.[28] Despite much sophisticated elaboration, Neuburger offers the essential myth as his description of Hippocrates. What Galen and the Renaissance writers believed were facts about Hippocrates' deeds and writings have

---

28. Sir T. Clifford Allbutt, in his *Greek Medicine in Rome* (London, 1921; repr. New York, 1970), offers a most thoroughgoing, and therefore unconvincing, elaboration of ancient medical history as a series of brilliant scientific beginnings by geniuses who were misunderstood by their followers.

now been removed as unverifiable. Only the spiritual essence remains.

LITTRÉ AND THE MODERN HIPPOCRATIC QUESTION

One might reasonably ask why historians who admired the outlook of science were not more skeptical about basic assumptions and more curious about the provenance of the questions they were asking about the Corpus Hippocraticum. I cannot answer that question adequately, but rather will take Emile Littré, the great editor and interpreter of Hippocrates, as both example and cause: he read Hippocrates in his own image and in the image of the medicine of his time, and, as the last complete interpreter for whom Hippocratism was alive and meaningful in day-to-day medical practice, he has largely determined the range and the course of subsequent discussion of the Hippocratic question.

Maximilien Paul Emile Littré (1801–1881), the physician philologist, did his monumental translation and interpretation of Hippocrates between 1839 and 1861. His stated purpose was to improve medical practice by making Hippocrates' works available to his fellow medical men in their own language.[29] At that time, medicine was about to undergo radical changes. Louis Pasteur published his *Mémoire sur les corpuscules organisés qui existent dans l'atmosphère* in 1861. Joseph Lister published his first important paper on antiseptic surgery in 1867. Throughout the first half of the nineteenth century, progress was made in mi-

29. Littré trained as a physician but did not practice. Besides the entries in various biographical dictionaries, one can read his autobiographical sketch at the beginning of his collection of papers on medical subjects, *Médecine et médecins* (Paris, 1872), pp. i–viii. Indicative of Littré's timeliness is what René Laennec, famous for his development of the stethoscope, wrote in 1804 in his *Propositions sur la doctrine d'Hippocrate relativement à la médecine pratique*. He hoped that a physician-philologist would come forward to reveal the systematic principles that guided Hippocrates. He himself offered some modest attempts in that direction. (See M. Martiny's summary in *La collection hippocratique,* pp. 97–105.)

croscopy and histology. Rudolf Virchow's book *Cellularpathologie* was published in 1858. In the latter part of the nineteenth century, Germany began to outdistance France in medical research and teaching. From 1815 to 1840, France, and particularly Paris, was the center of creative activity in medicine; and Hippocrates was a symbol of much of the creativity. One creative activity was expectative therapy, as opposed to active intervention. Between 1789 and 1844 the death rate at the hospital Hôtel Dieu was halved because of neglect of therapeutics,[30] largely by hygienic measures such as giving every patient his own bed and by avoiding excessive drugging and "vampirism" (venesection). Hippocratism was also associated with treating the patient, not the disease. Among physicians, according to Erwin Ackerknecht, "The notion of specific disease experienced an all-time low between 1820 and 1880."[31]

Littré studied Hippocrates, then, at the last time when Hippocrates was very relevant to current medicine. Littré's work, in that sense, will never be superseded or replaced, although his text and his historical facts have been improved. No current scholar believes Littré's outline of ancient Hippocratism in its entirety. But a comparison of W. H. S. Jones's Loeb Classical Library translation or of Hans Diller's German translation with that of Littré shows that Littré's work is still indispensable. Francis Adams, the physician counterpart of Littré in England, was commissioned by the Sydenham Society to translate "the genuine works of Hippocrates" while Littré's work was in progress.[32] He was heavily indebted to Littré and reinforced his influence in England and America. Comparison of modern books on Hippocrates (for example, William Arthur Heidel's

---

30. On the Paris hospital reforms, see Erwin H. Ackerknecht, *Medicine at the Paris Hospital, 1794–1848* (Baltimore, 1961), death rate tables, p. 17. I am much indebted to Ackerknecht's book for information about medicine and medical thought in France at the time Littré and Charles Daremberg were doing their work.

31. Erwin H. Ackerknecht, "Aspects of the History of Therapeutics," *Bulletin of the History of Medicine* 36 (1962), 408.

32. See Francis Adams, *The Genuine Works of Hippocrates* (London, 1849), Translator's Preface. His is the last translation in English that I know by an author who considers Hippocrates useful in day-to-day practice.

Hippocratic Medicine) with Littré will inevitably show that disagreements are few and shallow: the notion of "Hippocratic spirit" has remained as Littré left it.

His work is influential because it is the last of its kind as well as eloquent and complete. I wish to draw attention to the fact that its Hippocratism is to a great extent that of the Paris school. The notions still called up by "Hippocratism"—such as emphasis upon the patient, not the disease, and emphasis upon observation to the exclusion of theory—are notions promoted by Littré and his contemporaries, René Laennec, Antoine Laurent Bayle, and others, as the keys to medical progress.[33] Littré promoted Auguste Comte's positivism (he was a founder of the *Revue Positive* in 1855). His notion, therefore, of what medicine needed and what Hippocrates could offer was up to date.

Hippocrates, Littré said, used essentially the experimental method of modern science, differing from it primarily in that he had a smaller number of facts at his disposal (1.463).[34] Hippocrates' profound contribution was that he saw a disease as a single entity which develops through related stages. Lacking detailed knowledge of the seats of disease and of anatomical facts, Hippocrates turned his research to the common elements of diseases observable through external symptoms and concentrated on prognosis: prognosis distinguishes the science from empiricism and blind practical treatment (1.454). But Hippocrates' science is based on reality, in contrast to the philosophical medical systems of his time and to the medical systems that tried to replace his (1.462–463). These are the main points Littré makes eloquently and at length to illustrate the "rapprochement que l'esprit fait entre la science moderne et la science antique," the comparison the mind makes between modern and ancient science (1.xiv).

33. P. J. G. Cabanis spoke of Hippocrates as his own predecessor, the first sensualist: see Ackerknecht, *Medicine at the Paris Hospital,* p. 4.
34. Emile Littré, ed., *Oeuvres complètes d'Hippocrate,* 10 vols. (Paris, 1839–1881). The first volume contains the introductory essays explaining his principles and the text and translation of *Ancient Medicine.* Essays of varying lengths introduce the individual works in subsequent volumes. I shall refer to Littré's edition in the normal manner by volume and page numbers, for example, L 6.476: Littré's edition vol. 6, p. 476.

Littré contrasts the scientific outlook of Hippocrates with that of the Cnidians on the one side, who failed to construct any science but dealt with diseases individually, and on the other side with the philosophical systems that did not relate their speculations to reality (1.472–473).

Littré gives historical reality to the Hippocrates he imagines in an excursus that takes up more than 550 pages of his first volume. He seems to mention all suggestions by anyone in antiquity about Hippocrates' reality and importance, with little emphasis on the contradictions in ancient tradition and little emphatic skepticism about the quality of the evidence. In his beautifully arranged presentation, the tradition, properly read, all points in one direction. Most impressive is his account of each of the works of the Corpus in which he assesses the work's nearness to Hippocrates' style and points of view.

He took what he described as "the unanimous testimony of antiquity" in favor of some of the works as decisive for their Hippocratic authorship; hence he judged *Epidemics* 1 and 3, *Prognostic, Airs Waters Places, Regimen in Acute Diseases,* and some others as genuine. He found Hippocrates' theoretical pronouncements in the treatise *Ancient Medicine,* a judgment for which he could hardly claim antiquity's unanimous testimony, but modern Hippocratism had prepared the way for that decision. Littré simply turned Galen's arguments for *Nature of Man* around and stated that *Ancient Medicine*'s inductive method is what Plato describes as Hippocrates' method. To make historical sense of the condition of the works, Littré inferred that the Corpus as a whole was first assembled and published in Alexandria in the third century B.C. From the varied condition of the works, which range from garbled compendia and private notes to finished literary productions, he inferred that some had been previously published and circulated, but that most came to Alexandria from a family library, whence their publication in Alexandria fixed their texts for the future. The earliest commentators were confused about the authorship of many of the works, and that confusion persisted. But about some works there was no doubt; hence the "unanimous testimony" for them, which he found verified by his own judgment about their doctrine. Although he had to sift ancient testimony to arrive at his conclusions, Littré assumed that the information he sought had been stated clearly in books by Erotian and Galen on the genuine and false works of Hippocrates, but that these books accidentally and regrettably had been lost. He lamented their loss and also of the history of medicine produced in the fourth century B.C. by Aristotle's pupil, Menon. But, Littré reasoned, since Menon's authoritative history was available from the fourth century on, nobody could have made any very extreme statements about Hippocrates' doctrines in contradiction to Menon. His lament for the loss of Menon's work is especially interesting because of the later shock to the scholarly world when excerpts from Menon's work were discovered. Littré writes

> If the work had survived, or if Galen had discussed it in establishing the doctrines that are characteristic of Hippocrates, we would certainly have precise evidence about ancient medicine in general and Hippocrates in particular. A work as ancient as Menon's, Aristotle's pupil, would settle many questions about the date of this or that discovery, or this or that theory, would eliminate at once what is posterior to that philosopher, and would give us precise notions about the period between Hippocrates and the peripatetic school. The very topic of Menon's book would go directly to our purpose and furnish us with most precious material for a history of medicine before Aristotle, one of the periods when documents are most rare and uncertain. [1.167]

Littré's use of Galen is inconsistent, however: he infers that Galen knew the truth and therefore loss of some of Galen's writings deprives us of the truth. Yet he dismisses Galen's testimony about Hippocrates' scientific outlook, the authenticity of *Nature of Man* as compared with *Ancient Medicine,* and so on. Littré never clears up these and similar inconsistencies by discussing how Galen's evidence is to be used. No one has systematically done so since. Hence the subject matter of my Chapter 2.

Littré's picture of Hippocrates proved very congenial to students of ancient medicine and stimulated much further research. But his historical construct did not hold up. Attempts to achieve more precise descriptions of the history of the Hippo-

cratic texts produced blanks or baseless elaborations of Littré's
fantasies: questions about what Corpus Hippocraticum Diocles
had or what sorts of commentaries on Hippocratic works
Herophilus wrote produced such unsatisfactory answers that
eventually scholars had to infer that they were asking the wrong
questions. Between 1891 and 1930, new information, along with
more thorough investigation of material Littré had used, required scholars who knew more and more to admit to greater
and greater areas of ignorance. Their reluctance to go backward
is evident in their writings. Far from being burning skeptics, the
scholars who developed the Hippocratic Question show a will to
believe and a reluctant retreat in the face of the facts. One looks
in vain for a radical critique of the subject. Some influential
articles will serve for illustration.

In 1891, the prolific medical historian Max Wellmann wrote
confidently in Littré's manner about Alexandrian critical research into the authorship of works in the Corpus.[35] Shortly
afterward, to everyone's surprise and shock, fragments of Menon's lost history of medicine came to light in a papyrus roll
from Egypt. The papyrus was edited by Hermann Diels, historian of ancient philosophy and science and tireless organizer of
scholarly projects, including the Corpus Medicorum Graecorum
(CMG). The papyrus, which is called Anonymus Londinensis
because it is stored in the British Museum, was written in the
second century A.D., but contains excerpts from Menon's reports
of physicians' views about the causes of disease. Menon's report
of Hippocrates' doctrines was readily recognized by Diels to be a
digest of the doctrines of one of the works of the Corpus Hippocraticum, *Breaths* (*peri physôn*). But Diels was perplexed because
no knowledgeable student of the subject would take *Breaths* to be
work of Hippocrates. In agreement with Littré and the rest of
the scholarly world, Diels considered *Breaths* to be a wretched,
sophistical, pseudophilosophical exercise that was unrelated to
the school of Cos, let alone to the spirit of Hippocrates, author
of *Airs Waters Places* and careful observer of facts. I quote Diels's

35. Franz Susemihl, *Geschichte der griechischen Literatur in der Alexandrinerzeit*, I
(Leipzig, 1891), 778–779.

comparison: "Denn hier spricht ein zugleich geistvoller und nuchterner Arzt, ein philosophisch durchgebildeter, aber nicht mit dem Scheine philosophischer Bildung prunkender Forscher, ein geschmackvoller Stilist, aber kein poetisirender, archaisirender, antithisendrechselnder Sophist. Menon hat sich also in der Person des Hippokrates grundlich versehen."[36] (Here in the Hippocratic works speaks a physician who is both inspired and sober, one who was educated in philosophy, not one who offers specious research with a pretense of philosophical education; a tasteful stylist, not a poetizing archaizing antithesis-turning Sophist. Menon was basically mistaken about the identity of Hippocrates.) To explain the occurrence, Diels made up a story: once upon a time Aristotle put one of his pupils, Menon, to the task of composing a doxography of medical thought. When Menon looked for the work of Hippocrates to excerpt, he went to the Lyceum library and there passed over the genuine Hippocratic works that lay at hand, such as *Epidemics, Airs Waters Places,* and *Prognostic,* but he picked up the wretched work *Breaths,* which was stored with them. Menon was attracted to *Breaths* because he leaned to pneumatism, and *Breaths* seemed to support his views; therefore he created the report of Hippocrates given in the papyrus.

In the aftermath of the discovery of the Anonymus Londinensis papyrus, scholars found it easier to give up the possibility of attributing specific extant works to Hippocrates than to give up Littré's notion of Hippocrates the founder of scientific medicine. Friedrich Blass led the way in 1901, arguing that Menon's report of Hippocrates' doctrines was not really an accurate excerpt from *Breaths.* Probably, he conjectured, Hippocrates wrote a much better, more intellectually respectable work, which the author of *Breaths* knew and reflected in part. But Hippocrates' work, which Menon read, must have been lost subsequently.[37] In 1910, seventeen years after his article on the

---

36. Hermann Diels, "Ueber die Excerpte von Menons Iatrika," *Hermes* 28 (1893), 429; whole article, pp. 407-434. Diels asserts that it would be aesthetic sin and substantive error to accept *Breaths* as Hippocratic.

37. Friedrich Blass, "Die pseudhippokratische Schrift *Peri Physon* und der Anonymus Londinensis," *Hermes* 36 (1901), 405-410.

contents of the papyrus, Diels wrote of the skepticism that by
then reigned among researchers on the subject and cited a
suggestion by Wellmann that Hippocrates, like Socrates, had
written nothing, despite his great influence.[38] Wellmann reviewed the problem again in 1926: he conjectured that Menon
had made no error (he was, after all, trained in medicine and
knew his texts); rather, Menon must have attributed the doctrine
of *Breaths* to some other Hippocrates, perhaps Hippocrates'
grandson, the son of Thessalus, but that later excerpters of his
doxography dropped the patronymic, "son of Thessalus."[39]
Again in 1929, Wellmann summed up the grounds of skepticism
on the subject and proposed another unconvincing solution.[40]
The Corpus, he said, must have been catalogued by a careless
librarian at the Alexandrian Library, who indiscriminately
lumped the most discrepant works together under Hippocrates'
name and in the process effectively effaced any evidence for
attributing them correctly. There is no evidence that Alexandrian or post-Alexandrian critics had earlier commentaries or
sound testimony to rely on, and therefore their testimony about
authorship of works is worthless, whether or not it is unanimous,
because all their judgments are simply inferences from the Corpus itself. But Wellmann saw the glimmer of a possibility.

38. "Ueber einer neuer Versuch, die Echtheit einiger hippokratischen Schriften
nachzuweisen," *Sitzungsberichte der preussischen Akademie der Wissenschaften,*
philosophisch-historische Klasse (1910), pp. 1140–1155. In this article, Diels
contributes to the skepticism by arguing that fifth- and fourth-century authors did not cite and criticize specific passages in one another's works and
that, therefore, we cannot infer (as later antiquity did) that a fifth- or
fourth-century criticism of a medical procedure represented criticism of a
passage in a Hippocratic work. His argument is addressed to an attempt by
Hermann Schöne to prove *Joints* and *Fractures* genuine on grounds of such
criticisms of medical procedures.
39. Max Wellmann, "Hippokrates des Thessalos Sohn," *Hermes* 61 (1926),
329–334. Wellmann does not say what happened to Menon's report of Hippocrates himself. He was apparently still proceeding on the theory that
Hippocrates wrote nothing.
40. Max Wellmann, "Hippokrates des Herakleides Sohn," *Hermes* 64(1929),
16–21. Wellmann concludes his article with an interpretation of Plato's description of Hippocrates' method in the *Phaedrus:* perhaps by "knowledge of
the whole" Plato meant knowledge of all external factors that influence
health, which are certainly considered in *Prognostic* and *Epidemics* 1 and 3.
Karl Deichgräber and Max Pohlenz were later to take up the suggestion.

Herophilus, who preceded the careless cataloguing, was said to have criticized Hippocrates' *Prognostic,* a fact that would be witness of *Prognostic*'s genuineness. If *Prognostic* is genuine, so are *Epidemics* 1 and 3, whose material is apparently related to *Prognostic.*
Willy nilly, scholarship progressed by attenuation of confidence in the historical structure Littré had erected to support his view of Hippocratic science and what Hippocrates wrote, despite scholars' apparent lack of desire to attack Littré's opinions. Meanwhile, people set out to elaborate aspects of Littré's views, most notably his conception of the Cnidian school. Johannes Ilberg, in 1925, developed the idea of a coherent Cnidian school, parallel to and even influencing Hippocrates' Coan school. He looked for common authorship and similar mentality in various of the treatises Littré had judged Cnidian and developed the idea of a Cnidian school which had a library and an archive and whose works somehow got mingled with the Coan works (perhaps in a Coan library) and then transported to Alexandria, where they were all attributed to Hippocrates.[41] Ilberg's development of Littré's thesis certainly did not weaken it.
Ludwig Edelstein in the early 1930s came closest to a radical critique of Littré's views.[42] Edelstein considered the prominence of prognosis in works of the Corpus to be not incipient science, but response to social needs: the physician had to establish and protect his reputation in day-to-day dealings with patients and their families. In the absence of licensing or other credentials, the itinerant physician developed prognosis to a great degree for self-protection and display. Science, in a modern sense, it was not. Edelstein's principal text is *Airs Waters Places,* which de-

---

41. Johannes Ilberg, "Die Aerzteschule von Knidos," *Berichte über die Verhandlungen der sächsischen Akademie der Wissenschaften zu Leipzig,* Philologisch-historische Klasse, vol. 76, pt. 3 (1924).
42. Edelstein's views were first presented in *Peri Aeron und die Sammlung der hippokratischen Schriften* (Berlin, 1931), and developed in an article ("Nachträge") in Pauly-Wissowa-Kroll-Mittelhaus-Zeigler, *Realencyclopädie des klassischen Altertums* (Stuttgart, 1894———), Suppl. 6, cols. 1290–1345; herafter cited as *RE.* Parts of his book have been published in English translation in Ludwig Edelstein, *Ancient Medicine: Selected Papers,* ed. Owsei Temkin and C. Lilian Temkin (Baltimore, 1967).

scribes how a physician can anticipate the diseases of a population from their climate, soil, and water supply. Moving on to
other prognostic works of the Corpus, he found that the implicit
theories were inconsistent with *Airs Waters Places* and with each
other, that there was no unified theory or approach that can be
identified with a scientific movement which Hippocrates led.
Though Edelstein did not so describe his approach, it seems that
he offered a *reductio ad absurdum* of assumptions about "science"
that had been read into the Hippocratic Corpus. He then did the
same with the other arguments for Hippocratic authorship. He
surveyed the ancient testimony about the works and, turning
Littré's procedure upside down, found no work whose authenticity had *not* been questioned. He also accepted Blass's arguments that *Breaths* could not have been the source of Menon's
report of Hippocrates' doctrine. Yet he argued that, because
Alexandrian and post-Alexandrian testimony about Hippocrates was worthless because accurate information about Hippocrates did not reach Alexandria, we must take Plato's and Menon's reports about Hippocrates as definitive for our picture of
Hippocrates: since these reports describe no extant work, the
writing of Hippocrates must be presumed to be lost. It is clear,
Edelstein argued, that a myth about Hippocrates, father of
medicine, developed after the Alexandrian period, and that it is
based on the Corpus Hippocraticum, which cannot be considered the work of Hippocrates. All of the elements of Edelstein's
critique had been anticipated in previous scholarship, but his
determined arguments for a discontinuity between the early and
late conceptions of Hippocrates and for the futility of traditional
approaches appear to have been indigestible to most other
scholars in the field because they were depressing. Edelstein
observed, in an interesting aside, that *Regimen in Acute Diseases*
had excellent credentials as a Cnidian work because it criticized
and improved on an earlier Cnidian work, the *Cnidian Opinions,*
but answers to his suggestion only reasserted its similarity to
works thought to be Coan.[43] Edelstein nailed his thesis on the

43. For Edelstein's suggestion see *Peri Aeron,* p. 154. For the replies, see Iain M.
    Lonie, "The Hippocratic Treatise *peri diaites oxeon,*" *Sudhoffs Archiv* 49
    (1965), 50–79.

door of German scholarship in an unsettled time and later defended it from the United States. To charges that his skepticism was robbing us of our Hippocratic heritage, he responded that the truth was worthwhile and that several other great men's works were also lost.[44] The more traditional position was reasserted by Karl Deichgräber in an influential book published in 1933, *Die Epidemien und das Corpus Hippocraticum,* another outline of the history of the Coan school of medicine along the lines that Littré had laid out, with corrections to allow for intervening scholarship.[45] Deichgräber renounced skepticism about Hippocrates' relation to the Corpus Hippocraticum and tried to make a plausible case for the old view: some of the works could be by Hippocrates and/or his students because they are apparently related to one another in their doctrine and could very well have been written in the right period. Accepting that assumption, he argued that three generations of the Coan school were represented in the three groups of books in the *Epidemics*: 1 and 3 from the time of Hippocrates, ca. 410 B.C., are characterized by humoral pathology and attention to prognosis, and they share the numerical system of *Prognostic,* hence they must be by the same author. *Sacred Disease* and *Airs Waters Places* are probably from a different author, though from the same period. *Epidemics* 2, 4, and 6 can be dated to ca. 395 B.C., and they show influence both of philosophical notions and sophistical rhetoric and of Herodicus of Selymbria, who had specific ideas about regimen and who is mentioned unfavorably in *Epidemics* 6. These works can also be related to the surgical works of the Corpus and to *Humors* and *Nature of Man.* The third group,

44. Ludwig Edelstein, "The Genuine Works of Hippocrates," *Bulletin of the History of Medicine* 7 (1939), 247–248. This is the conclusion of an article surveying recent work in the subject.
45. *Abhandlungen der preussischen Akademie der Wissenschaften,* philosophisch-historische Klasse (1933), no. 3. My page numbers in the text refer to this work. Cf. Edelstein's response to this work (note 42), and that of Henry Sigerist, "On Hippocrates," *Bulletin of the History of Medicine* 2 (1934), 209. For a recent assertion of the validity and usefulness of Deichgräber's work see Hans Diller, "Stand und Aufgaben der Hippokratesforschung," *Jahrbuch der Akademie der Wissenschaften und Literatur* (Mainz, 1959), 271–287, esp. 278–281.

Epidemics 5 and 7, Deichgräber dated to ca. 360. In surveying the ancient testimonies about Hippocrates and his students and family, Deichgräber found some that he could trust despite the rampant skepticism: he was impressed by a late genealogy that traced Hippocrates' family back to Ascelpius and by lists of Hippocrates' students, some of which could have been valid (*Epidemien,* pp. 147–149). He saw no reason to mistrust a statement in the Suda (a lexicon of the tenth century A.D.) that a student of Hippocrates, Dexippus, ended a war against Cos by healing the sons of the Carian king Hecatomnus because there is no counterevidence (pp. 166–167). In renouncing skepticism, Deichgräber effectively eschewed the consideration of alternatives. "Und doch ist diese Skepsis, so behaupte ich, im Grunde nicht berechtigt. Wenn die Philologie den Masstab anlegt, nach dem sich ihr Urteil zu richten hat, dann muss sie zugeben dass ihre Skepsis unberechtigt oder voreilig ist" (p. 7). In effect, the position of Littré must be maintained as long as there is the slightest possibility of supporting it. Deichgräber seems to have been right in sensing that the nineteenth-century confidence that scholarship would advance step by step to truth had been exhausted.

Max Pohlenz, in his book on Hippocrates, rejoiced that one was free again to believe and to talk about the personality of Hippocrates, founder of scientific medicine.[46] W. H. S. Jones, who worked on the relationship between philosophy and medicine, said, "The question of authorship is not likely ever to be settled. On the other hand, we do possess the *Corpus,* of which several books are in the true sense great achievements, with consistent doctrines inspired by all that is best in the scientific spirit. The Hippocratic problem, like the Homeric problem, cannot take from us our heritage."[47]

Interest and activity in the subject declined somewhat after the mid-1930s, but there has been a striking revival since the

46. Max Pohlenz, *Hippokrates und die Begründung der wissenschaftlichen Medizin* (Berlin, 1938), pp. 1–2.
47. *The Medical Writings of Anonymus Londinensis,* trans. William H. S. Jones (Cambridge, 1947), p. 20. Cf. his article, "Hippocrates and the Corpus Hippocraticum," *Proceedings of the British Academy* 31 (1948), 1–23.

1950s. Renewed progress in the Corpus Medicorum Graecorum and in other series of better texts and translations has been accompanied by a rise in the number of interpretive studies.[48] Examples are colloquia held at Strasbourg in 1972 and Mons in 1975 to bring together scholars from many countries who are working on aspects of the Hippocratic Corpus and its role in the history of medicine.[49] Besides demonstrating the liveliness of current activity, the papers presented at the conferences seem to indicate a modern consensus as to what the interesting problems are and how they are to be solved. Recurrent themes for discussion show that the conceptual basis for current work on Hippocratica is that of Littré as it was refined in the late nineteenth and early twentieth centuries: for example, the sorts of scientific and technical thought that characterized the schools of Cos and Cnidos; the chronologies of the "schools"; the primitive elements that remain in the works. Hippocrates, his personality, and his scientific ideology are missing from current discussion: he is now an unknown quantity in his school.

The history so far covered in this chapter suggests compellingly that thought about Hippocrates and the Corpus had followed to the end the channel created for it by the Enlightenment rediscoverers of Hippocrates, who embraced him as the inspirational ancient counterpart of the new science. Readers of the Hippocratic works have been betrayed by their very enthusiasm for some texts of the Corpus into staying within that channel and accepting the premises of Le Clerc or Sprengel. Evidence and interpretive possibilities have been ignored or only partially used because of the way in which the subject has been conceived. There is good reason to reexamine the whole of the evidence systematically and critically. One must ask anew of the Corpus in antiquity where it came from and how, what works people read,

48. See Diller's account of recent work, note 45 above. The interpretations of Jacques Jouanna, Iain Lonie, Fridolf Kudlien, Robert Joly, Charles Lichtenthaeler, Hermann Grensemann, and Louis Bourgey listed in the bibliography offer a good sample of current work.

49. *La collection hippocratique;* see note 5 above. *Corpus Hippocraticum; Actes du colloque hippocratique de Mons* 22–26 septembre 1975, ed. Robert Joly. Editions Universitaire de Mons, série sciences humaines IV (Mons, 1977).

and how they conceived them at each stage, what sort of Hippocrates they imagined from their reading, and how that imagined Hippocrates influenced the medicine that was practiced, and vice versa, what sorts of anachronistic readings of ideas into Hippocrates are related to the medical practices of each period. For the study of those questions in antiquity Galen's evidence is crucial, since it has survived in such quantity and spread its influence over subsequent ages.

One preliminary question can be approached, however, while virtually ignoring Galen's evidence: the question whether Hippocrates wrote anything, and if so, what. It has been successfully demonstrated by scholars in the nineteenth and early twentieth centuries and conclusively argued by Edelstein that the tradition divides into two, and that the question what Hippocrates wrote must be argued on the basis of the pre-Alexandrian evidence of Plato and Menon.

Freed from many traditional questions about the Corpus, I find that the pre-Alexandrian evidence about Hippocrates along with later vague traditions that he invented or perfected regimen lead rather convincingly to a conclusion about his writing which is hitherto unheard-of.

HIPPOCRATES, AUTHOR OF *REGIMEN*

Plato's allusive discussion of Hippocrates' doctrine and method in the *Phaedrus* is agreed to be the earliest testimony, although it is simply a by-blow in the discussion of oratory, which is the subject of the *Phaedrus*. To the question what makes a great orator, Plato's answer is, of course, philosophy, the only true science, which teaches one what he must know to pursue the other sciences or crafts (*technai*). Socrates wittily proves that point to the young Phaedrus by example, by argument, and by analogies. Medicine is a *techne* like oratory, which can be practiced well or badly depending on whether or not one practices it philosophically. Plato's purpose in using Hippocrates as the example of the philosophical physician has been variously interpreted and with varying degrees of literalness in the reading of Plato's words.[50]

Socrates' argument, in the relevant part of the *Phaedrus*, is that to be a great orator one must not follow the path (*methodos*) of Lysias and Thrasymachus, but that of Pericles (269d),[51] that is, the path Pericles followed when he associated himself with Anaxagoras, who filled him with *meteorologia* (i.e., lofty thoughts) and drew his attention to the nature of mind and mindlessness, a subject on which Pericles spoke a great deal (270a). At this point, Socrates brings up medicine: "The case with the craft of medicine is perhaps the same as that of oratory." "How do you mean?" says Phaedrus. "In both," says Socrates, "you must analyze a nature: of the body in one and the soul in the other. If, that is, you are going to use not only practice and experience, but *techne*. In the one you communicate health and strength to the body by means of drugs and nourishment; in the other you communicate to the soul excellence and persuasiveness by administering the customary discourses." Phaedrus responds, "That is probably true." "Well, then," Socrates replies, "do you think that you can learn anything worthwhile about the nature of the soul without learning the nature of the whole?" Phaedrus answers "If one believes Hippocrates the Asclepiad, not even of the body without that method." This is Plato's first indirect description of Hippocrates' doctrine.

Phaedrus' elliptical answer to Socrates' question seems to be an acknowledgment of Socrates' allusion to a sentiment expressed by Hippocrates, such as, "One cannot learn about the nature of the body without learning the nature of the whole." All interpreters agree on this point. But the sentiment itself has caused

---

50. The scholarly literature on the subject is immense, and I will not try to cite it in detail. No one has suggested what I am suggesting. In "La question hippocratique et le témoignage de Phèdre," *Revue des Etudes Grecques* 74 (1961), 69–92, Robert Joly gives a lively review of literature and opinions on the subject. He divides scholars into those who think Plato could have been speaking of extant antiphilosophical Coan works and those who do not.

51. Plato, *Phaedrus*, ed. John Burnett, Oxford Classical Texts (Oxford, 1901).

much discussion because of its built-in ambiguity: does "nature of the whole" mean "nature of the whole man" or "nature of the whole cosmos"?[52] I see no reason to think that Plato does not intentionally suggest both, and perhaps a third meaning as well, "the nature of all body." As always, Plato uses his language precisely and self-consciously, and if he is purposely ambiguous he will take account of all the meanings he has suggested.

After the first sentiment about method is attributed to Hippocrates, the dialogue continues to elaborate it. Socrates answers, "Yes, an admirable statement. Still, we must scrutinize the *logos* alongside Hippocrates and see whether it agrees." "Yes," says Phaedrus. "See, then," says Socrates, "what Hippocrates and the true *logos* say on the subject of nature. Must we not think about a thing's nature thus: first, is the thing in which we want to be craftsmen (*technitai*) and capable of making others craftsmen simple or multiform? And second, if it is simple, we must consider its power (*dynamis*): what power has it by nature for being acted on by what? If it is multiform, we enumerate the forms and observe of each form what we observe of the one: by what it naturally does what, or by what it is naturally acted on and how." Phaedrus answers, "I suspect that you are right, Socrates." At this point in the dialogue, Socrates drops the analogy with medicine and discusses how oratory will use the method of Hippocrates that has been outlined. But what is that method?

The passage quoted above contains several allusions back to the earlier part of the *Phaedrus* that are relevant to its interpretation. "True *logos*" alludes to Socrates' "true" speech earlier in the dialogue, which he offered in contrast to a pair of false speeches. The true *logos* (*logos* meaning both "speech" and "line of reasoning") proceeds by definitions and distinctions which Socrates described as follows: the speaker of the true *logos* surveys and brings together into one *idea* (form or concept) the things he wants to teach, so that he can give a single definition (265d), and

52. A review of recent scholarly work on the subject was given by Harold Cherniss, "Plato, 1950–1957," *Lustrum* 4 (1959), 139–141. Cherniss inclines to the view that the "whole" cannot mean the universe.

then he again separates them by *eidê* (types, forms), dividing them by their natural joints and not breaking any *meros* (part) like a bad butcher (265e). Socrates calls these techniques *synagogê* and *diairesis* (collection and division) (266b). Lysias and Thrasymachus (examples of people who claim to be *technitai* in oratory) know nothing of these techniques (266c). When, in the passage discussed above, Socrates alludes to the true *logos,* he suggests that Hippocrates uses collection and division in his *methodos.* With his use of the word *technitai,* Socrates alludes to his previous discussion of the term in which he uses physicians as examples of those who either are or are not *technitai*: if a man went to the physician Eryximachus, or his father Akoumenos, and said that he could apply drugs to make people warm or cold and to make them move their bowels or vomit and that therefore he was a physician and could make other people physicians, they would think him insane. The man lacks *technê:* he has experience, but he does not know whom to treat, when, and with what quantity (268a–c).

Here is the description of method in *Regimen* which I believe Plato uses in the *Phaedrus*:

> I contend that whoever is going to write properly about regimen for men must first know and distinguish the nature of man as a whole. He must know (*gnônai*) from what things man is composed from the beginning, and must distinguish (*diagnônai*) the parts (*merê*) by which he is controlled. For if one does not know the original composition he cannot know what results from those things. And if he does not know what is to control the body, he cannot know how to administer what will benefit a man. These things the writer on regimen must know, and next what power (*dynamis*) each food and drink in our regimen has by nature or by human constraint and *technê.* [Bk. 1. ch. 2, L 6.468]

One must know exercises and

> how to proportion exercise to the bulk of the food, to the nature of the man, to the ages of the bodies, to the seasons of the year, to the changes of the winds, to the situations of the regions in

which the patients reside, and to the constitution of the year. A man must observe the risings and the settings of the stars so that he can guard against changes and excesses of food, drink, and winds, and of the whole cosmos, from which things diseases come for man.

Not even that is enough, continues Hippocrates; he himself has discovered, in addition to all of the foregoing factors, *prodiagnosis*: the technique of finding out by symptoms what kind of excess exists that will likely produce disease, so that the disease can be prevented.

This impressive outline of a science of medicine is, I contend, what Plato refers to in the *Phaedrus*. A closer comparison of what Plato says with what Hippocrates says clarifies Plato's reasons for using Hippocrates' work as he does in the context he does. Plato sees a parallel between Hippocrates' *gnosis* and *diagnosis* (which mean "know together" and "know separately"; I translated "know and distinguish") and his own collection and division. Hippocrates says that one must know the nature of man as a whole and must know the parts that control him and the *dynamis* of all aspects of the environment that affect man. Plato found this a useful parallel to his thoughts about scientific oratory: distinguish kinds of souls and how each acts or is acted on and classify speeches according to their *dynameis* in relation to particular souls (*Phaedrus* 271a–b). Plato left "the whole" ambiguous because both "man as a whole" and "cosmos" are comprehended in the Hippocratic science. I suspect also that Plato made a pleasant pun with "true *logos*," referring by it not only to his own earlier speech, but to Heraclitus as the inspiration for the philosophical Hippocrates (as Anaxagoras was for Pericles). The *logos* of Hippocrates is a description of all things in flux, all part of a process. In chapter 4 of book 1 of *Regimen* (L 6.476), he says, "Whenever I speak of becoming or perishing I am merely using popular expressions. I really mean mingling and separating. The truth is, becoming and perishing are the same thing, mixture and separation are the same thing. . . . Yet nothing of all things is the same." Reminiscence of Heraclitus is unmistakable.

Heraclitus also said that while the *logos* is common, people do not comprehend it, but act as though they are deaf.[53] After outlining Hippocrates' method in the *Phaedrus,* Plato says that people who don't comprehend it are like blind and deaf men (*Phaedrus* 270e). One would infer that Plato saw the connection between Hippocrates and Heraclitus.

Plato's procedure in the *Phaedrus* can be better understood by comparing what he says with the work he refers to, a confirmation of the identification. We can go further, perhaps. Plato's example of the operations a man would learn before (falsely) claiming to be a physician were to make people warm or cold and to make them vomit or move the bowels. Why these examples? *Regimen* addresses itself to heat and cold as operative principles in diet, drugs, and environment. All must be balanced. If evidence of imbalance is seen and the physician suspects *plesmone* (too-muchness), he flushes out the patient, fore and aft, with emetics and cathartics (which should not overheat), reduces the diet, and prescribes the right exercises and foods for warming and cooling until he gradually brings the patient back to the right condition. Simple remedies, as Plato says, but the secret is to know when, to whom, why, and in what amounts. Plato reflects not only the philosophical theories of *Regimen,* but its medical substance as well.

We cannot tell what Plato really thought of Hippocrates. Plato says that Hippocrates is to medicine as Pericles is to oratory, but elsewhere Plato offers serious strictures about Pericles' effect on the souls of Athenians. In the *Phaedrus,* his reference to Pericles' talk of "mind and mindlessness" is an ironic allusion to Pericles' dependence on Anaxagoras' philosophy of mind (*nous*), about which Plato is elsewhere less than complimentary. The same irony may be present in his treatment of Hippocrates' philosophic medicine and his true *logos.* Plato uses Hippocrates for his urbane purposes to say that if method is needed for dealing with bodies, it is much more needed for dealing with

53. Heraclitus, fragment B 34, in Hermann Diels, *Die Fragmente der Vorsokratiker,* 10th ed. by Walter Kranz (Berlin, 1952), I, 159.

souls. Galen's enthusiasms were required, as we shall see in Chapter 2, to turn Plato's urbanity into a confession that he derived his method and his major doctrines from Hippocrates.

We have, then, discovered Hippocrates' work which Plato cites, and we can see why Plato uses it as he does. As Plato might say, the answer has been there all along. Why have scholars not tumbled to it? As I suggested above, students of the subject have failed to look at what was there because they had set for themselves the impossible task of finding "science" like Galen's, or like Bacon's. Many years ago I noticed the startling similarity between the method asserted in *Regimen* and Plato's report of Hippocrates, as well as the relation between Menon's report of Hippocrates and *Regimen*. I made a note of it and forgot it. It could only become significant in the light of a survey of the Hippocratic tradition. I propose it here somewhat as Hippocrates proposed *prodiagnosis* to the medical profession: it puts the subject on a new basis.

Regimen is an attractive, individualistic, and literate work, written in the mature tradition of classical dietetic medicine (it acknowledges its predecessors in that field and says that it will add to them), which tries to give a comprehensive view of man's relation to his environment in health and disease. But it is based on the sort of wrongheaded hypotheses that have offended people who held an ideal of inductive science. Its explicit theories also offended Galen, who held to a four-humor theory and looked for a Hippocrates in his own image. I will suggest that Menon, the first known historian of medicine, read the work and based his report of Hippocrates' doctrine on it and, in the process, shaded his report of Hippocrates' theories to match his own, whether he idealized Hippocrates because Plato had spoken of him as he did or because the reputation of Hippocrates was already considerable in the fourth century. Menon's use of *Regimen* is not as obvious as Plato's, but in the end I find it equally convincing.

The extant fragments of Menon's history of medicine, although unlike the lost work Littré imagined and uncongenial to Diels, do tell us what ideas Menon had of the development of medical theory and what he made of Hippocrates.[54] The surviving excerpts are organized according to causes of diseases that Menon's predecessors had recognized, reduced to the two essential ones: *perissomata* (excrement, primarily what is left over from processes of digestion) or *stoicheia* (the elemental components of the body). Within each category Menon put each physician or philosopher whose views he reported into a sequence in which each disagrees in some particular with those who precede, so that the doxography is presented as a dramatic progression, an intellectual dialogue reminiscent of Aristotle's discussions in his philosophical works of his predecessors' views. Hippocrates is third among those who thought that diseases come from *perissomata*. Here is the sequence.

Euryphon said that "when the belly (*koilia*) does not discharge the nutriment that has been taken, *perissomata* are produced, which rise to the regions about the head and cause diseases. When, however, the belly is empty and clean (*lepte kai kathara*) digestion takes place as it should. Otherwise, what I have already said occurs" (4.31–40). Herodicus, whom Menon next reports, partly agreed and partly disagreed, according to Menon's account. He agreed that *perissomata* cause disease, but disagreed that it is due to the condition of the *koilia,* whether it is *kathara* and *lepte*. Herodicus, says Menon, explains: "If one takes food without exercising, it is not absorbed, but lies in the belly plain and unaltered and is dissolved into *perissomata*. From the *perissomata* come two liquids, one acid and one bitter, and the affections differ according to the dominance of one or the other. And he says that according to the strength or weakness of these, the affections resulting are different: what I mean is, if the acid is rather weak and not unmixed, and analogously if the bitter be not too bitter, but somewhat less, or if they are strong, the affec-

54. Diels's edition of Anonymus Londinensis was published in *Supplementum Aristotelicum,* 3, pt. 1 (Berlin, 1893). W. H. S. Jones published the text with an English translation and essays in 1947 (see note 47 above). Jones, however, does not indicate how much of the text that he prints is conjectural restoration. All interpretations must be checked against the excellent original publication of Diels.

tions will differ according to the degree of mixture of the liquids" (5.6–21). From there the report wanders on for twelve more lines to say that the affections differ also according to the loci of the liquids.

These first two reports provide the context for the views of Hippocrates that follow. Menon does not quote the people whose doctrines he reports, but translates them into the language of his own categories. What specific words of Euryphon and Herodicus, if any, Menon is reporting with his own word *perissomata* is not clear. Menon's dramatic sequence, in which Herodicus "disagrees" with Euryphon, produces an absurdity, whereby Herodicus' view seems to be that it does not matter whether the *koilia* is *kathara* and *lepte*, but it does matter whether one exercises. The report of Menon (but here it could be the fault of the anonymous source who excerpted from him) shows a tendency to maundering repetitiveness on points that fit the dramatic sequence of cumulative doctrine, such as the strength of humors and their location, while otherwise the account seems brief and sketchy.

Hippocrates' peculiarity, according to Menon, was to say that gas (*physai*) causes disease. I quote Menon's report, and also passages from *Regimen* and from *Breaths* (*peri physon*), for comparison. *Regimen* seems to be the source of Menon's report, although Diels and others have taken *Breaths* to be the source.

Anonymus Londinensis 5.35–6.13

(5.35) But Hippocrates says that the cause of disease is gas (*physai*), as Aristotle reports him. For Hippocrates says that diseases are brought about in the following fashion: either because of the quantity (*plethos*) of things taken, or the diversity (*poikilia*), or because they are strong and hard to digest, *perissomata* are produced.

(5.44) And when the things taken are too many, the heat that effects digestion is overcome by so much food, and does not effect digestion. And because it is hindered, *perissomata* are produced.

(6.4) And when the things taken are varied, they quarrel (*stasiazei*) among themselves in the belly, and from the quarrel comes change into *perissomata*.

(6.7) When foods taken are difficult to digest, there is hindrance of the digestion because of the difficulty of digestion, and thus a change into *perissomata*.
(6.11) And from the *perissomata*, gas rises up, and the gas arising brings on diseases.

Breaths, Chapter 7 (L 6.98–100)

This is bad regimen, when one gives more wet or dry food to the body than it can bear, and opposes no labor to the quantity of nourishment. Also when one ingests foods that are varied (*poikilas*) and dissimilar. Dissimilar things quarrel (*stasiazei*), and some are digested faster, some slower. Along with much food, much air (*pneuma*) must also enter, since along with everything that is eaten or drunk, more or less *pneuma* enters the body. That is proved by the following: people belch after much food and drink, since the air (*aêr*) rushes up when it breaks the bubbles that contain it. When the body is full of food it becomes full of *pneuma* when the foods remain long. The food remains because its quantity keeps it from passing through. When the lower intestine is blocked the gas (*physai*) rushes through the whole body and falls on the parts that are full of blood and chills them.

Regimen

(75,L 6.616, cf. Anon 5.44) There also occurs the following kind of *plesmone:* the next day food is belched up raw but not acid. . . . In this case the belly is cold and cannot digest the food in the night.[55]
(56,L 570, cf. Anon. 6.4) Meats in sauces cause burning and water, since fat, fiery and warm foods which have powers opposite to one another are residing together.

55. *Regimen in Health* (the last part of *Nature of Man*) offers a parallel to this description: "Those who vomit up their food on the next day, and whose hypochondria are swollen because of the undigested food, should sleep more and be subjected to fatigue, and should drink wine in greater quantities with less water and at the same time eat less food. It is clear that their belly cannot digest the excess (*plethos*) of food because of weakness and coldness" (L 6.84). This must have been standard etiological doctrine in dietetic works, emphasized by some more than by others.

(56,L 568-570, cf. Anon. 6.7) Raw foods will cause intestinal pain and belching, since what fire should do must be done by the belly, which is too weak for the ingested foods.

(74,L 614-616, cf. Anon. 6.11) There also occurs this kind of *plesmone:* when the food digests in the belly but the flesh does not receive it, the nourishment stays and makes gas (*physa*). [When this increases for some days] the remnant in the belly overpowers whatever is ingested, raises the temperature, disturbs the whole body, and produces diarrhea (followed by dysentery).

Although *Breaths* does not have the etiological theory Menon attributes to Hippocrates—that diseases are caused by gas which is produced from *perissomata* from altered food—*Breaths* does have the not dissimilar notion that gas ingested with food is trapped by hindrance of digestion and that it causes disease by blowing through and chilling the body. *Regimen,* on the other hand, has precisely the theory Menon attributes to Hippocrates: altered food that stays in the intestines creates gas that causes disease. (*Regimen* is, to my knowledge, the only work in the Corpus Hippocraticum that actually expresses this theory.) *Regimen* also has all the elements of Menon's threefold scheme for the causes of the problems of digestion, whereas *Breaths* has only two. Exercise is important in both *Regimen* and *Breaths*, but not in Menon's report of Hippocrates, though Menon mentioned that factor earlier in his report of Herodicus and did not need to repeat it. Neither *Breaths* nor *Regimen* would have produced the report Menon gives if his method had been simply to excerpt directly. Either treatise could have produced his report if he was making his own scheme and sequence for his doxography so that it would lead from one writer on regimen to the next, with a climax at Hippocrates. *Regimen* seems to me to be the best candidate as Menon's source. And, I submit, with Plato's evidence, the chances seem excellent.

After he reports Hippocrates' doctrine, Menon pauses to dramatize it and to ask whence Hippocrates could have got his view. His conjecture is somewhat absurd because it does not relate directly to the etiology of disease that he has attributed to Hippocrates:

Hippocrates said these things because he was moved by the following conviction: *pneuma* is the most necessary and most powerful component in us, since health comes from its free flow and disease from its impeded flow. We are like plants: as they are rooted in the earth, so we are rooted in the air by our nostrils and whole body. We are like those plants that are called soldiers; as they, rooted in water, are carried now here, now there, so we, like plants, are rooted in air, and go now to one place, now to another. Since that is so, when *perissomata* occur, *physai* arise from them, and rising up like mist (*anathymiatheisai*), cause disease. And from the difference of *physai* diseases are produced. If they are many they cause disease, and again if they are few, they cause disease. And from change of *physai* diseases come. They change in two ways, toward excessive heat and excessive cold. Whichever way the change comes, it produces disease. This is what Aristotle thinks about Hippocrates. [Anon. Lond., ch. 6]

Menon's scheme and the conjectures he offers can be understood only by comparing what he says with some of Artistotle's statements about such matters. Not only does Menon seem to have imported the common Aristotelian word and concept *perissomata* into the text of the people he reports, but he seems to have read into Hippocrates' *physai* from unassimilated food the Aristotelian notion of *anathymiaseis,* fumes, which account for sleep and for diseases in Aristotle's scheme; and he seems to have tossed in the metaphor "rooted in air" as an elaboration of Aristotle's related notion that the head in animals corresponds to the root in plants (*On the Soul* 416a.2–5; *Parts of Animals* 686b.34–35; *Progression of Animals* 705a.26–b.8. Aristotle may have been inspired by Plato's conceit that man is a celestial plant, his head like a root attached to the heavens [*Timaeus* 90a]). Aristotle spoke frequently of the *anathymiaseis* from food. Most useful for comparison with Menon is his explanation in *Parts of Animals* (652b.33–653a.10) of fluxes: nourishment evaporates, and its fumes travel upward through the blood vessels; its *perissoma,* cooled by the coldness of the brain, condenses and causes flows of phlegm and serum, which cause inflammation and disease. Aristotle compares the process to rain showers and adds that insofar as it relates to natural philosophy he will speak of it in the *Origin of Diseases,* a book that has not been preserved. I suggest that Menon made up a line of reasoning for Hippocrates in a manner similar to that of the master himself, Aristotle, who, for example, made up a Thales who thought in Aristotelian categories: Thales, according to Aristotle, said that the principle of all things is water, wherefore he showed that the earth is on water. "And he probably derived this supposition from seeing that nutriment is wet, and the heat itself comes from it and exists by it, since what things come from is the principle of all things" (*Metaphysics* 1.5, 983b.21–26). The author of the papyrus that contains the excerpts from Menon rejects Menon's account of Hippocrates, substitutes his own account of what Hippocrates said, and then criticizes what Hippocrates said as wrong. The first part of this author's report of Hippocrates' doctrine is badly mutilated in the papyrus; the second part, which he criticizes, is apparently from *Nature of Man*: epidemic diseases come from the atmosphere, whereas sporadic diseases are caused by individual regimen. The author does not give the origin of Menon's version, or even of his own. Menon appears to have used *Nature of Man* as the basis for his report of Polybus' medical views (in chapter 19 of Anon. Lond., which is fairly badly multilated).

That Menon distorts the material he excerpts in the direction of his own ideas can be demonstrated from his report of Plato's theories of the causes of diseases, which is based on the *Timaeus*. Plato gets considerably more attention than anyone else in the excerpts from Menon that we have (170 lines, compared with 24 for Hippocrates). In them Plato's elemental theories are reported in Menon's own language and with considerable rearrangement, followed by his descriptions of three causes of diseases: the first from excess or displacement of elemental components, the second and third from dissolution of tissues.[56] For the example of Menon's tendencies, I translate Menon's report

56. Lists of passages in the *Timaeus* on which the author of the papyrus draws, with notes on his method of rearranging them and changing the language, are offered by Diels and Jones.

of Plato's third cause of disease beside my own paraphrase of the passage in the *Timaeus*.

Timaeus 84d–85c	*Anonymus Londinensis* 17–18

The third kind of disease one must conceive in three ways, as it comes from *pneuma*, from phlegm, and from bile.
  1. *Pneuma:* when the lungs are blocked by fluxes (*rheumata*), the air enters one part in excess, while unable to enter the blocked portions. The portions deprived of *pneuma* rot, the portions distended by it suffer pains and illness. *Pneuma* which is produced by dissolved flesh, and which cannot get out, causes the same pains (as the above), the worst of which involve the sinews and veins (the diseases tetanus and opisthotonus).
  2. White phlegm is dangerous when it is trapped inside because of bubbles of wind in it. If it can get out it causes eruptions on the exterior of the body; if it mixes with black bile it attacks the sacred nature of the brain (epilepsy).
  3. From bile come all inflammations. When bile finds an outlet it causes eruptions. Trapped inside, it causes many kinds of burning diseases, the worst of which comes when bile overpowers the congealing properties of the blood.
	Diseases come from *perissomata* in three ways: either from gases (*physai*) from the *perissomata* or from bile or from phlegm. On account of these three, in common or separately, diseases come. That is to say, one of them brings disease, and two occurring together bring disease, and similarly diseases are brought about by the three of them when they are joined together.

Nobody could recover the inner logic of Plato's system by reading Menon's account. (The papyrus is in very good condition here, so that the translation above can be accepted with confidence.) Menon has not only altered the vocabulary and emphases, he has assimilated Plato's scheme to his own scheme of *perissomata*. Yet the *Timaeus* unmistakably inspired Menon's report.

While there is no straightforward solution to the problem of the origin of Menon's report of Hippocrates, one set of probabilities seems much more likely than another. We can assume that Menon accurately excerpted a lost work (we would be in the mixed company of Edelstein and Deichgräber, who drew different conclusions from that assumption), but we would have to acknowledge that no one in antiquity, no one before the twentieth century, had any inkling that Menon excerpted a work that was not afterward available to read. People who could read Menon entire did not notice that he excerpted a work they could not read, whereas Blass and all who followed him found it fairly obvious from the available excerpts. Modern scholars probably have misread Menon in assuming that he attempted to answer the questions being asked at the time his excerpts were discovered. Littré's earlier fantasy about what kind of medical history Menon would have written is not irrelevant. One can also fantasize about the purpose and method of a work, or fragments of a work, that is before one's eyes. We have good grounds for asserting, in opposition to scholars since Blass, that Menon probably did not excerpt a work now lost. Rather, he probably reported a work in the tradition of dietetic medical writings, he organized his report to dramatize a peculiarity in Hippocrates' work which he found congenial to his own concepts, and he produced his own conjecture about how Hippocrates could have arrived at it.

This supposition accuses him of a kind of historical naiveté not uncommon in his time and acknowledges that he did not approach the subject as the nineteenth century did. It is appropriate to point out that no one in antiquity adduced Menon as a witness for the authenticity or lack of it of any work in the Hippocratic Corpus, but that is how moderns have tried to use him. The anonymous source of the papyrus' reports of Menon simply states that Menon has missed the point and then tells what Hippocrates really says. He does not say that Menon mistook Hippocrates' genuine work, nor does he say something like "it is suspected that Hippocrates' genuine work is lost, since Menon's description does not correspond with any extant work." This, despite the existence of a Hippocratic Question in antiquity for some centuries before the papyrus was written. The anonymous writer apparently treated Menon as another reader of the Corpus Hippocraticum who did not get the point.

I think it very likely, then, that Hippocrates wrote the work *Regimen* that has come down to us in the Hippocratic Corpus. How was the knowledge lost? Here are two descriptions of the work's pedigree from Galen:

> Those who think that spelt groats were not used in the time of Hippocrates are refuted by mentions of them in the ancient comic poets, and by Hippocrates himself in *Regimen in Health.* Even if that work is not by Hippocrates, but by Euryphon or Phaon or Philistion or Ariston, or some other ancient (they attribute it to many), still all those men are ancient, some older than Hippocrates and some his contemporaries. But I mean Hippocrates the son of Heraclides, whose book it actually is. His grandfather, Hippocrates the son of Gnosidicus, wrote nothing according to some people, and only the two books *Fractures* and *Joints* according to others. [CMG 5.9.1, p. 135]

> At this time it is time to speak of the *krasis* of foods, as it says in the book on *Regimen,* ascribed to Hippocrates by some, and also to Philistion, Euryphon, and Philetus, all men of antiquity. It begins thus in some copies: "This is the way to distinguish the *dynamis* of each food and drink, both its natural one and that imparted by *techne*" [the beginning of our present chapter 39 in Book 2], but in others it begins thus: "The position of and nature of each place you should distinguish thus" [the opening of our present Book 2]. When this book is transmitted by itself it is called *Regimen,* being the second part of the whole which is divided into three parts. But when the whole is found undivided, composed of three parts, it is entitled *On the Nature of Man and Regimen.* The second part, which talks about foods, one might properly think worthy of Hippocrates. The first part departs entirely from Hippocrates' view. But let this be said as an aside on the way. To whatever of the men named it belongs, it seems to refer regimen in food to a general method. [CMG 5.4.2, pp. 212–213]

Many such asides by Galen give us our history of the Hippocratic Corpus, which I shall pursue below. These two give an inkling of the history of Hippocrates' work. From the time of the collection of ancient medical literature under Hippocrates' name in the Alexandrian Library, *Regimen* was only one of many works transmitted under Hippocrates' name. Its useful section on properties of foods was sometimes transmitted separately from the theoretical introduction and the *prodiagnosis,* and its titles varied. Hippocrates dated quickly, like other mortals, and some people preferred other works in the Corpus to his and even conjectured who else might have written it. Galen was inconsistent, depending on what he was attentive to at the moment. When he was attentive to the theoretical aspects of *Regimen,* Galen could not accept it because he read Hippocrates in his own image and found the four-humor theory of *Nature of Man* more congenial. Modern scholars have responded to *Regimen* with both praise and depreciation. It has recently been edited, correctly dated, in my view, and placed in the tradition of dietetic medicine by Robert Joly.[57]

So much for a brief preliminary sketch of the history of what I think is Hippocrates' own work. The history of *Regimen* is only part of the history of the Corpus, of ideas about Hippocrates in antiquity, and of the ways medical men oriented themselves to Hippocrates. I shall begin the consideration of the larger story with Galen. His information is crucial, and one must search it out in the immense body of his works and then learn how to interpret it. For all his influence, Galen is not well known to us, particularly in works in English. We can usefully pursue Galen's intellectual biography in relation to his ideas about Hippocrates, while characterizing the information he offers in his writings about the Hippocratic tradition before his time.

57. *Hippocrate, Du Régime,* ed. and trans. Robert Joly (Paris, 1967). An excellent article on the structure and import of the work by Hans Diller is "Der innere Zusammenhang der hipp. Schrift *de victu,*" *Hermes* 87 (1959), 39–56. Also useful is Joly's *Recherches sur le traité pseudo-hippocratique du Régime* (Paris, 1960).

2

GALEN'S HIPPOCRATISM

In examining Galen's Hippocratism and whatever is pertinent
to it, I shall try to answer these questions: where is he original
and what did he inherit; what effect on his medicine did his
Hippocratism have; were his views consistent throughout, or did
they change and develop; and most importantly, what is the
quality of the evidence he offers for the reconstruction of the
preceding tradition? I shall follow his career and writings
chronologically, insofar as possible, in order to see what sort of
intellectual biography of Galen's career in relation to his Hippocratism can be constructed.[1] I come to Galen as a classicist, not

1. The basic, indispensable study of the chronology of Galen's life and works is
the four articles by Johannes Ilberg, which I cite as Ilberg 1, 2, etc.: "Ueber
die Schriftstellerei des Klaudios Galenos," *Rheinisches Museum für Philologie*:
(1) n.s., 44 (1889), 207–329, (2) 47 (1892), 489–514, (3) 51 (1896), 165–196,
(4) 52 (1897), 591–623. They were supplemented and in part corrected by
Kurt Bardong, "Beiträge zur Hippokrates-und Galenforschung," *Nachrichten
von der Akademie der Wissenschaften in Göttingen*, phil.-hist. Klasse, Nr. 7 (1942),
pp. 577–640 (cited as Bardong). I shall state my own few disagreements with or
corrections of the conclusions of Ilberg and Bardong. Otherwise I shall not
notice divergent views save where they seem important. Two recent noteworthy studies relate specifically to Galen's orientation toward Hippocrates:
Georg Harig and Jutta Kollesch, "Galen und Hippokrates," in *La collection
Hippocratique et son rôle dans l'histoire de la médecine* (Leiden, 1975), pp. 257–274;
and Loris Premuda, "Il magistero d'Ippocrate nell' interpretazione critica e
nel pensiero filosofico di Galeno," *Annali dell' Università di Ferrara*, n.s., sec. 1
(1954), 67–92. Both these articles attempt to evaluate Galen's claims to be
Hippocrates' follower and interpreter. Neither follows chronologically the
development of Galen's Hippocratism. References in the text to Galen's writings are to *Claudii Galeni Opera Omnia*, ed. Carolus Gottlob Kühn, 20 vols. in
22 (Leipzig, 1821–1833), abbreviated K, or to volumes of the Corpus

as a medical man. At the best I hope to read him as a Greek or Roman would have read him in his own day, while adding historical perspective.

EARLY MEDICAL TRAINING

Galen always expressed adulation of and gratitude to his father. Nikon, as Galen recalls him, exhibited perfect self-control, unfailing kindness, and incisive intellectuality. He was impatient with opinions that could not be substantiated, and he deprecated adherence to "movements" (that is, sects, *haireseis*). An architect, Nikon held mathematics as his model for thought, and he cautioned against emotional views that could not be submitted to demonstration with logical precision. Galen retained an image of his mother as a person governed by emotion, subject to irrational passion and extreme anger (CMG 5.4.1.1, 27-30).

Living in Pergamum in Asia Minor, and perhaps more intensely Greek for that, Nikon saw that his son was educated in Greek language, the classical authors, and rhetoric and philosophy, especially dialectic. The training seems to have been directed toward making Galen a professional philosopher and teacher (a sophist), which was an educational ideal of the period. But the god Asclepius, whose great temple in Pergamum was then being constructed (perhaps with the assistance of Nikon himself), intervened and called Galen to his service by means of a dream sent to his father. At sixteen, therefore, Galen was apprenticed to a physician who was companion to Costunius Rufus, who was responsible for constructing Asclepius' temple (K 12.224-225). Galen had fifty more years of life in which to live up to his special call from Asclepius, and he did well.

Nikon lived only about five years after Galen began his medical and philosophical studies in Pergamum, and on his death left

Medicorum Graecorum, various editors, published by the Academia Berolinensis Hauniensis Lipsiensis, 1907-present, abbreviated CMG, or to the three volumes of *Scripta minora*, ed. Ioannes Marquardt, Iwan Müller, and Georg Helmreich (Leipzig, 1884-1893), abbreviated *Scr. Min.* References to other texts are specified.

Galen money enough to be comfortable through his life. After his father's death, Galen pursued his medical training in Smyrna, Corinth, and Alexandria before returning, at the age of thirty, to Pergamum to assume the apparently prestigious position of physician to the gladiators attached to the temple. Two years later, he moved to Rome, where within four years he achieved a considerable reputation for his healing and for his writing on medical and philosophical subjects. After a brief return to Pergamum, he went back to Rome in 169 A.D. at the age of forty and for the next quarter of a century wrote prolifically on medical and philosophical subjects, while serving as physician to the imperial court.

Although he pursued his medicine as a result of the call from Asclepius, Galen's training and practice were thoroughly secular. He did not, like some physicians, act as a servant to the god in carrying out the prescriptions patients received in dreams, nor did he discuss and evaluate temple medicine in his works. We do know of one relevant incident, through a description in the *Sacred Tales* of Aelius Aristides, in which the rational and irrational met: Satyrus, Galen's first teacher, visited the famous hypochondriac sophist Aristides and offered him an ointment for his serious skin condition and also advised him that he had lost too much blood from phlebotomy. Aristides, who had put himself in the god's care and was following the treatment the god prescribed in his dreams, accepted the ointment but did not use it immediately. The god had not recommended it. When he used it some time later, he believed that he was punished by catching consumption for doing so and ever after clove to the god's commands and to physicians who would carry them out. Aristides speaks of Satyrus as an eminent sophist.[2]

With that "eminent sophist," Galen began his anatomical and medical studies in Pergamum, which he continued for approximately eleven years under teachers whom he sought out, meanwhile pursuing his philosophical studies. He became the most accomplished anatomist of his time, or probably of any time

2. Aelius Aristides, *Sacred Tales* 3.8–9. Galen did himself have dreams about Asclepius and on the god's advice opened an artery in his hand, with salubrious effect (K 11.314.19).

before Vesalius, though he did not dissect humans. And he became an interpreter of the "ancient medicine," whose principles he claimed to have brought to perfection as none had before him. Unlike the modern medical student, Galen had to choose the philosophical basis of his science from diverse competing theories. He made and refined his choice gradually and not without influence from his teachers.

At the risk of some repetition later, when we deal with sources of Galen's views as expressed in his writings, I think it will be useful to give here a brief account of Galen's teachers and their contributions to his education in Hippocratic science, to attempt, that is, to put together the hints in Galen's writings about the environment from which he came. The composite picture of his teachers' views must be constructed from his scattered references to them in one hundred or so treatises written over some fifty years, though the bulk of his explicit references to them comes from the last part of his career. The evidence he gives is consistent enough to permit confident inferences.

Galen's final pronouncement on the "orthodoxy" of contemporary writings about Hippocrates is from his work on the *Categories of His Own Books:*

> If I die before I explain the most significant elements in Hippocrates' writing, those who want to know his view will have my treatises, as I have said. Together with the commentaries I have written they should read these other commentators: my teacher Pelops, and the books of Numesianus (few have been preserved), and also those of Sabinus and Rufus of Ephesus. Quintus and his students do not properly understand Hippocrates' view, and hence often go wrong in explanations. Lycus sometimes even criticizes him and says that he is mistaken, though Lycus does not know his *dogmata.* [Yet the books of Lycus have been praised.][3] But my teacher Satyrus (I studied with him before Pelops) did not give the same explanations as Lycus of the Hippocratic books. And Satyrus is agreed to preserve most accurately the views of Quintus without adding or taking away. Aephicianus leans toward Stoicism. But I, having heard Satyrus' explanations of Quin-

3. The text is deficient in this sentence, but the bracketed sentiment is probably what it said.

tus, and later having read some of Lycus' works, condemned both
as not knowing accurately Hippocrates' views. Those who followed Sabinus and Rufus knew better. He who is trained well in
my efforts can judge and detect what they say well and whether
they err.   [*Scr. Min.* 2.86–87]

Except for Rufus, all those mentioned are members of the medical community that produced Galen (see Figure 1). Lycus was
not Galen's teacher, though he was one of Quintus' students.
Distaste for Lycus precipitated, or crystallized, Galen's disaffec-
tion for Quintus as a master of Hippocratic interpretation.
Lycus' sin was reading the *Epidemics* as though they were written
by an Empiric; that is, reading the *Epidemics* as though the
statements in them were derived from experience and observation, not from an implicit philosophical theory. Lycus used the
Coan Prenotions and *Prorrhetic* as authoritative texts to explain
the *Epidemics*. Quintus' sin may not have been so extreme, but he
was not orthodox by Galen's final judgment. Galen's views
changed and developed after his years as student, but they
began as his own peculiar combination of the approaches to
medicine and to Hippocrates that came from his teachers. In
time, Galen's talent for verbal abuse was exercised against his
teachers as well as against virtually everyone else, and single
statements must be balanced against the whole picture. Here I
shall attempt a synthetic sketch of Galen's immediate predecessors, beginning with Quintus' teacher, Marinus.

Marinus was very much admired by Galen for his researches
in anatomy, which revived and examined the work of
Herophilus and Erasistratus four centuries earlier. Marinus was
very significant to Galen's view of Hippocrates in one respect: he
was the first to assert that a Hippocratic work (*Epidemics* 2) contained a correct anatomy of veins and nerves.[4] The anatomy is
extremely obscure, to be sure, and very brief, but it can be rec-

4. Marinus' views about Hippocrates' anatomy of veins and nerves are spoken of
in the commentary on *Epidemics* 2 (CMG 5.10.1, pp. 309–313, 330–331).
When a student, Galen condensed Marinus' twenty anatomical books to four
as an exercise: CMG 5.10.2.2, p. 377; *Scr. Min.* 2.104; *On Anatomical Procedures: The Later Books*, trans. W. L. H. Duckworth (Cambridge, 1962), pp.
9–10.

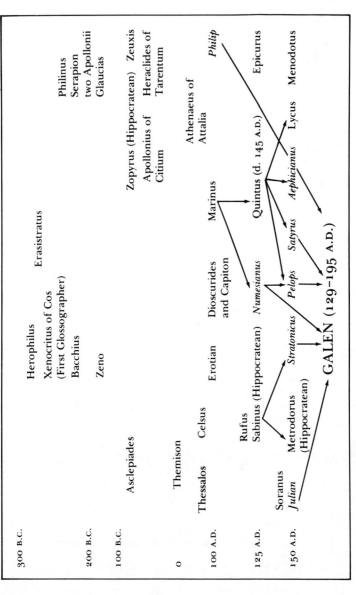

Figure 1. The people on this chart are those who figure most prominently in Galen's history of the Hippocratic tradition in Alexandria. People are placed on the time chart by rough floruits. Specific teacher-student relationships are indicated by arrows. Galen's teachers are italicized.

onciled with the truth as Herophilus had described it in his
anatomy. Marinus had confirmed its accuracy in his own dissections of apes. Other descriptions of blood vessels in works of the
Hippocratic Corpus, says Galen, are all mistaken. Hence this one
correct anatomy is very important to Galen's argument that
Hippocrates knew everything and never erred. What else
Marinus said about Hippocrates and in what works is not clear.
At one time Galen implies that he wrote many commentaries on
Hippocrates which were readily available, but wrong. The
statement occurs in one of Galen's typically snide polemical
asides, and, as often in such cases, it is difficult to judge whether
Galen was simply carried away by the desire to say something
insulting. The statement occurs in the commentary on *Epidemics*
6 (CMG 5.10.2.2, pp. 286–288): Lycus commented on a reading
of the text which no one else appears to know, and, after quoting
Lycus' comment, Galen remarks that he did not meet Lycus, who
had no great reputation during his own lifetime, though after
his death people admired his books for their clear explanations.
Galen read them, he says, and found that they followed Marinus
in all respects, save that they were more long-winded. Therefore, says Galen, he searched the shops for Marinus' book to see
how he had written the passage. But, says Galen, he could not
find it, though there were many of Marinus' books in Rome. I
am suspicious of this statement by Galen because he never mentions a Hippocratic commentary by Marinus otherwise—he cites
him for no readings or explanations, nor does he mention him
in the passage quoted above on Hippocratic authorities, nor in
the list of commentators whose works he had read and made
extracts from in his student days (CMG 5.10.2.2, pp. 412–413).
Galen seems to have enjoyed putting Lycus down and deprecat-
ing his anatomical works, which were much admired in Rome,
by saying that Lycus took everything from Marinus but, even so,
failed to understand him properly (*Anatomical Procedures* bk. 14.
ch. 1, Duckworth, pp. 184–185; *Anatomy of Muscles* K 18B 926–
927). I suspect that, for polemical effect, he transferred this
judgment to Lycus' Hippocratic commentaries. Hence, Marinus
may or may not have written commentaries on Hippocratic
works. Marinus' great importance for Galen remains the fact

that he revived the languishing tradition of Alexandrian anatomy by dissecting apes and other animals and passed his work on to Quintus (CMG 5.10.1, p. 312).

Galen repeatedly says that Quintus wrote nothing, yet Galen cites him for some few new readings and interpretations in the *Epidemics* (CMG 5.10.2.2, pp. 314, 500; CMG 5.10.1, p. 222), material which probably was transmitted to Galen by Satyrus, who faithfully reported Quintus' sayings (CMG 5.10.2.2, p. 412; K 14.69). Hence it appears that Quintus was an admired authority in Hippocratic interpretation as well as in anatomy in Galen's school years. Galen sought out Quintus' disciples to get as much of his teachings as he could at secondhand. Quintus was the Socrates of second-century medicine—an admired authority, whose disciples ultimately disagreed about the implications of his oral teaching. Undoubtedly, Quintus solemnized the wedding between the "ancient medicine" and anatomy and promoted study of anatomy through actual dissection, for which all his students were known. But his view of the "ancient medicine" was so eclectic that he produced among his students an Erasistratean, Martialius; a Stoic, Aephicianus; quasi-Empirics, Lycus and Satyrus; and, through Numesianus, Pelops, who perhaps can only be called Hippocratean. That they were all, except perhaps Martialius, "Hippocrateans" (that is, they were not avowedly Empirics, Herophileans, Methodists, Pneumatics, or of any other sect), seems clear from Galen's calling Lycus a "bastard of the Hippocratic sect" because he leaned to Empiricism in his interpretations of Hippocrates (CMG 5.10.2.1, p. 13). Galen never would admit that Lycus was right about the thrust of Quintus' teachings, but he finally acknowledged, as in the passage quoted above, that Quintus' Hippocratism and his own were incompatible.[5] Galen inherited Quintus' insistence on pre-

5. I stress the indefiniteness of Quintus' doctrine because Johannes Mewaldt, "Galenos über echte und unechte Hippokratika," *Hermes* 44 (1909), 111–134, in the only extensive discussion of the question to date, inferred, erroneously, I think, that Quintus inherited and passed on to Galen an Alexandrian tradition of Hippocratic exegesis. In fact, no such neat tradition seems to exist. Wellmann's alternative assertion that the whole tradition of Hippocratic exegesis came to Galen through Sabinus will be discussed below in connection with Galen's commentaries. Harig and Kollesch, "Galen und Hippokrates," pp. 72–73, appear to agree with Mewaldt.

cise anatomical investigation as the center of medicine, but his mature views about Hippocrates came from elsewhere.

Numesianus was thought by Galen to be the best anatomist of his time. Galen followed him to Corinth and Alexandria to study with him (*Anat. Proc.* bk. 1 ch. 1, K 2.217–218). Unlike Quintus, Numesianus wrote about anatomy, but little of his work was published. He was secretive about some of his discoveries. After Numesianus' death in Alexandria, Galen cultivated his son so he could see Numesianus' books, but the son kept them to himself and was rumored to have burned them before his own death (*Anat. Proc.* 14.1). Galen says little about Numesianus' views about Hippocrates, citing him for the interpretation of only one passage, in *Epidemics* 2 (CMG 5.10.1, pp. 345–350). Numesianus' view is wrong, according to Galen, and Galen knew of the view not from the man's writings, but from Pelops' report of him. Galen considered that Pelops' view of the passage was also wrong, since he believed that one could infer a person's *crasis*, his temperament, from observation of his external features. Numesianus' views on the sources of blood vessels and nerves are not cited by Galen, but he probably thought he was wrong as were Pelops and everyone else. I infer that Galen recommends Numesianus' Hippocratism out of general piety for the man and because he brackets him with the Hippocratic Pelops, the best of Numesianus' students.

Pelops, whom Galen seems always to treat with great affection and respect as "my teacher," must have influenced him considerably in his early career, though Galen in time "corrected" Pelops' views about humoral pathology (including physiognomy and temperaments) and about anatomy of muscles, blood vessels, and nerves. Pelops' introduction to medicine was called *Introduction to Hippocrates.* Its third book contained the anatomy Galen corrected and its second the erroneous description of the sources of nerves and blood vessels. What else it contained and how it equated Hippocrates with medicine is unclear. Pelops also wrote private commentaries on Hippocrates' works, some few of which were published after his death. Galen read them all and excerpted them in his student days and in his later life used his notes in compiling his own commentaries. He said, "My teacher Pelops took great pains to alter obscure expressions and to find

explanations for them. . . . Since his explanations are always as
brief as possible, one cannot, as with others, point out his errors
and show that he has nothing useful to say." This grudging
compliment was written shortly before Galen recommended
Pelops' orthodoxy.[6]

Aephicianus and Satyrus, Quintus' other two students who
were Galen's teachers, left less distinct impressions. Galen rem-
inisces about Satyrus' instruction in anatomy (*Anat. Proc.* 1.14)
and quotes his report of Quintus' witty sayings (K 14.69). Satyrus
also wrote private commentaries on Hippocrates which Galen
read and excerpted (CMG 5.10.2.2, p. 412); they probably were
the source of Quintus' few views about passages in Hippocrates
which Galen later cited. Galen cites Satyrus himself only for a
view on a passage in *Prorrhetic*. Galen's purpose in that commentary was to debunk *Prorrhetic* and the tradition of Quintus that
used it, and Galen's comment on Satyrus' view is that it is
"neither demonstrable nor useful for prognosis" (CMG 5.9.2, p.
20). Aephicianus did not share Quintus' sin of Empiricism in
Hippocratic interpretation, but "leaned to Stoicism." Galen describes his reading of the opening of the Hippocratic work *The
Surgery,* which he interpreted as saying the same things as Simias
the Stoic's epistemological theory (K 18 B.654). How elaborately
Aephicianus worked out his Stoic interpretations of Hippocrates
is difficult to tell. I suspect that Aephicianus is the teacher who
introduced Galen to pneumatic elemental theory when Galen
was nineteen, and whom Galen later refuted.

Thus the medical tradition Galen inherited from Quintus was
focused on precise anatomical study as the basis for physiology,
and those who shared the tradition exhibited a wide range of
Hippocratic interpretation. From Sabinus and his student
Stratonicus Galen appears to have inherited a devotion to clinical medicine (bedside practice), and to Hippocrates as its source.
Stratonicus was the best of Galen's teachers in practical medicine

6. CMG 5.10.2.2, p. 291. For Pelops' humoral pathology, CMG 5.4.1.1., p. 75,
CMG 5.10.2.2, p. 500; physiognomy, CMG 5.10.2.2, pp. 345–350; the unpub-
lished commentaries, CMG 5.10.2.2, p. 412; his *Introduction to Hippocrates,
Anat. of Muscles*, K 18B 927, Iwan von Müller, ed., *Opinions of Hippocrates and
Plato* (Leipzig, 1874), 1.533–534.

(CMG 5.10.2.2, pp. 303, 412). Sabinus called himself a Hippocratean. He published his commentaries on Hippocrates and was
"famous" in Galen's time for his Hippocratic interpretations
(CMG 5.10.2.2, p. 510). In accord with Galen's habit of citing
people's opinions primarily to ridicule them or prove them
wrong, he treats Sabinus severely in casual remarks about him,
yet he approves of Sabinus' orthodoxy in his final assessment
quoted above. Galen was clearly in competition with Sabinus'
views during his own career, and in his works drew a picture of a
more complete, respectable, and scientific Hippocrates than he
had inherited from Sabinus.

Galen says that Sabinus "rashly set out to interpret Hippocrates without a dream of anatomy" and without experience of
dissection (CMG 5.9.1, p. 75; 5.10.1, p. 329). But he did so in
proper philosophical fashion: instead of considering the com-
parative accuracy of the different descriptions of blood vessels
and nerves in Hippocratic texts, Sabinus praised everything and
talked of the usefulness of vessels and nerves in the body. People
said that Sabinus was more accurate than previous interpreters
of Hippocrates and praised his explanations as "more clear than
Aristotle," so Galen tells us (CMG 5.10.2.1, p. 17; 5.10.1, pp.
329-30), but Galen cites him tendentiously, for talking nonsense, describing things that do not exist, and praising things
that are false. For his tendency to find significance even in such
details as the addresses of the patients in the *Epidemics,* Galen
manages amused tolerance (cf. CMG 5.10.2.1, pp. 11, 17, 167).
Sometimes Galen bursts forth with reasons why he could never,
like Sabinus, call himself a Hippocratean. He relates the story of
Philistion, a student of Metrodorus, who was Sabinus' student:
Philistion followed literally a recipe in *Epidemics* 2 for the cure of
barrenness. He served hot, half-roasted polyp (inkfish) to a lady
of good family, after demanding an immense fee. The lady
choked down two bites before becoming violently nauseated and
fainting. Philistion lost his patient and his reputation (CMG
5.10.2.1, pp. 401/402). The Hippocrateans in Alexandria had a
justification for the recipe: the polyp hangs on the rocks with its
suction cups (cotyledons) as the womb hangs onto the fetus. The
polyp is only slightly cooked so that its *pneuma* will retain its

quality and will influence the *pneuma* of the patient. But Galen says this is all nonsense. He expresses his disgust for people who call themselves followers of Hippocrates and who accept as true anything that goes under his name (CMG 5.10.2.1, p. 375). Teachers who pass such stuff off on students should be severely punished, he says.

Thus, though he finally recommended Sabinus' works when he listed people besides himself who were worth reading, Galen had no desire to be called Sabinus' follower or to place himself among the Hippocrateans. In a book attacking another of his teachers, Julian the Methodist, under whom he sat briefly in Alexandria, however, Galen defended the concept of Hippocratic science expressed in Sabinus' commentary on the *Aphorisms,* which Julian had attacked. Galen's work, based on his reading of a small part of Julian's, which he "hears is in forty-eight books," is more abusive against Julian than informative about the contents of Sabinus' work or Julian's criticisms. But one infers at least that Sabinus talked about imitating Nature in treatment by purging the noxious humors, whose plethora is the coherent cause of disease (CMG 5.10.3, pp. 39, 47, 52–53, 58–59). Julian, like Methodists generally, denied such causes and apparently such piety about nature.

Such is the picture Galen leaves of the instructors in medicine whom he acknowledged. Before proceeding to consider Galen's career and writings I shall offer some remarks about what studying Hippocrates under a teacher seems to have meant to Galen and how his own possessiveness on the subject of Hippocrates appears to me. Galen remarks that Julian and Thessalos (the founder of the Methodist school) could not have gone so wrong if they had read Hippocrates under a teacher (K 10.8; cf. CMG 5.10.3, p. 36). It was not enough to read Hippocratic works as medical textbooks. Hippocrates needed interpretation, like a proverb or a religious text, whereon, from a brief, pregnant statement, one can construct lengthy sermons and whole philosophies of life or of medicine. Students committed particularly pregnant statements to memory, such as "opposites cure opposites." Galen never could remember whence that statement came. He frequently attributed it to *Aphorisms,* though it comes from *Breaths*. From it the whole of medical theory can be inferred, especially if one remembers that hot is the opposite of cold, wet of dry. Like people who "know the Lord's will" in the most ambiguous situations, Galen knew to a certainty what Hippocrates would have thought about any subject, indeed did think, whether or not he had said so explicitly. Galen described his method in this way: "It is very characteristic of the method to be able to proceed from a brief elemental principle to the whole, part by part, and to judge anything that is erroneously said against a canon, so to speak, by comparing incorrect statements to scientific insights" (CMG 5.4.2, p. 53). He is not talking about logical testing of a hypothesis, but about interpreting Hippocrates, in this case a brief statement about massage. But for Galen logic, philosophy, and scientific method became one with Hippocrates and Hippocratic method. My comparison of his reading of Hippocrates with readings of religious parables is not casual. Galen had more fury for Lycus, the heretic, than for the infidels who had never been exposed to the sacred dogma.

He who is trained in Hippocrates can apply an aphoristic principle from *Aphorisms, Prognostic,* or the like to explain the facts described in therapeutic works, especially in the *Epidemics.* "He died on the seventh day." Of course, because "jaundice kills in seven days." "What would Hippocrates do in this case?" "Bloodletting, first day, of course, because 'therapeutic measures at the beginning. . . .'" This is typical of Galen's commentaries and his teacher's private commentaries and probably of many class sessions in Galen's training. The mystery game and scholarly wit were also included. The aphorism, "Growing things most heat," may seem to mean what it says, but it does not. If you feel babies and adults they are much the same temperature. But the secret is that the aphorism refers to heat which is and is not heat: the innate heat in the heart that makes us grow and function (on this subject Galen thought Lycus and others went dangerously astray).

Studying Hippocrates under a teacher meant, in part, then, introduction to an esoteric method and a manner of exegesis whereby the brief, bare, and even cryptic Hippocratic works could be read to yield the substance of modern medicine. The approach is unhistorical and antihistorical, and not everyone in Galen's period shared it equally, but it accounts for Galen's possessiveness toward Hippocrates. How much of that attitude he got from his teachers of medicine, how much from his philosophical studies, and how much from his own nature or temperament (as he might say) is difficult to judge, but I suggest that from his favorite teachers Galen adopted the posture of one who "Hippocratizes" by adducing apt quotes and illustrations from the Hippocratic Corpus, and who, in effect, tests the correctness of his views by his ability to find them in Hippocrates. As Galen made his own amalgam and advanced beyond his teachers, however, his "Hippocrates" had to develop with him, to the point that he was defending some outlandish interpretations. His passion was fed by the presence of boors and clowns in the medical and teaching professions who offended his sensibilities. He recalls one teacher (unnamed) who interpreted the symptoms in the case of Silenus in *Epidemics* 1, saying, "Naturally he was restless, talked, and laughed. He was a Silenus." The other students laughed and applauded, but Galen was deeply angry (CMG 5.10.2.1, p. 12). He condemns Julian the Methodist as an unserious member of an unserious sect.

I proceed to Galen's career and his writings. Two early treatises by Galen offer the basis for only modest inferences about his early Hippocratism. Galen began to be prolific early in his career, writing for friends. Of his earliest works his abstract of Marinus' *Anatomy* and his treatises on eye diseases and on thorax and lung movements are lost, but a brief treatise *On the Anatomy of the Uterus,* which he dedicated to a midwife,[7] remains, along with a treatise *On Medical Experience,* which addresses itself to the question whether Dogmatics, Empirics, or Methodists

7. I disagree with Ilberg's view that the treatise is too advanced to be given to a midwife and therefore must be a later rewriting. That the dedicatee was the only one for whose eyes it was intended seems unlikely (cf. Ilberg 2, pp. 490–491). References to the work in the text are to Kuhn 2.887–908. It is available in CMG 5.2.1, *Ueber die Anatomie der Gebarmutter,* ed. and trans. with notes by Diethard Nickel (Berlin, 1971). There is an English translation by Charles M. Goss in *Anatomical Record* 144 (1962), 77–84.

have the best philosophy of medicine. *Anatomy of the Uterus* was written in his student days in Pergamum before he was twenty-one (*Scr. Min.* 2.97). It is a very respectable work, based on his own investigation of apes and other animals (K 2.895). In it, his tone in approaching predecessors is somewhat more reverent than in *Anatomical Procedures,* which was written in his maturity. Yet already Galen challenges everyone except the divine Hippocrates, whom he praises, rather irrelevantly, for his description of the spine (p. 888), but whose ignorance he judiciously ignores in his discussion of fallopian tubes. Galen describes the fallopian tubes as seminal ducts and adds,

> Neither Aristotle, nor Herophilus, nor Euryphon knew these insertions. I mention these men not only because they were ignorant but because they are the best anatomists. It is not unexpected that Diocles, Praxagoras, Philistion, and practically all other ancients were ignorant of these as of many other parts of the body. They were general and not precise in anatomy, and I am not concerned with them. As for these others, I do not know what to say. I am not so bold as to condemn them, because of their accuracy in other matters, nor are these vessels so small that one would not notice them. [Pp. 900–901]

Later in the treatise Galen writes, "They say that there are no cotyledons in the human womb. They say that they occur in cows, goats, deer, and other such animals: damp, mucous bodies like the plant cotyledon, whence their name." But, he continues, Hippocrates does mention them, and so do Diocles and Praxagoras. How could they be wrong? Galen concludes that they are not wrong, but that they refer to the mouths of blood vessels, as Praxagoras explicitly says (pp. 904–906).

Thus, in Galen's earliest known work, his precociousness and his judicious and selective interpretation of Hippocrates are in evidence, already pointing toward his later claim that Hippocrates was preeminent and correct in anatomy as in everything else. His work *On Medical Experience,* written at about the same time, describes itself as a literary exercise in defense of experience in medicine against the attack of Asclepiades, the dogmatic theorist (ch. 1).[8] What he writes is not his own view, but those of Asclepiades and of the Empirics, who responded. He states elsewhere that the work was inspired by a debate between Pelops, not yet his teacher, and Philip, an Empiric (*Scr. Min.* 2.97). But the form of the work is not simply dramatic dialogue; Galen intrudes himself to evaluate the argument as it proceeds. After the Dogmatic speaker has argued the worthlessness of experience, because no two events are ever the same, Galen comments on the absurdity of the dogmatic position (ch. 8) before he gives the Empiric answer. The Empiric not only has the best of the argument, but, interestingly, he argues from Hippocrates and the Hippocratic Corpus. The *Epidemics,* repositories of experience that serve the memory and subserve gradual development in medicine, are evidence that medicine is based on experience, not *logos* ("reasoning," ch. 10). Hippocrates is quoted as saying, "Experience is necessary in food and drink" (ch. 13). Further, the Hippocratic view of critical days in disease is presented as pure empiricism: such a scheme follows no mathematical pattern and must be derived from experience (ch. 21). Very much in Galen's later style, a list of great physicians is offered (by the Empiric): not only Hippocrates, but Diogenes, Diocles, Praxagoras, Philistion, and Erasistratus all knew that knowledge can come only from reasoning from experience (ch. 13). Part of the list is repeated (again by the Empiric) to express the Galenic sentiment: only weakness and failure of reasoning lead men to be sectarians and partisans of one man, whether Erasistratus, Praxagoras, Asclepiades, Herophilus, or Hippocrates (ch. 24).

These expressions of Galen's own views and interests through the Empiric seem significant and seem to accord with the impressions he has left of tendencies in his early training under

8. *On Medical Experience,* ed. and trans. Richard Walzer (Oxford, 1944). The original Greek version has been lost, but the complete work survived in Arabic translation. Chapter numbers from his edition are cited in the text. I take the empiric tendency of the work to be more indicative of Galen's outlook than Walzer did in his introduction. Cf. Ludwig Edelstein's review of Walzer, *Philosophical Review* 56 (1947), 215–220, esp. 216. Edelstein would date the work late, but I do not agree with his reasons.

Satyrus. The treatise *On Sects for Beginners,* written somewhat later, in Rome, shows a later stage of Galen's self-definition. In that work, which is in dialogue form, the Empiric and Dogmatic together demolish the Methodist, while treating one another with respect.[9] Later in his career Galen showed less tolerance for Empiric arguments and for their claims on Hippocrates, probably in reaction to the immediate social and medical situation he confronted when he went to Rome.

GALEN'S FIRST RESIDENCE IN ROME

Rome? What can I do there? I don't know how to lie.
Praising a bad book and begging for a copy
Is far beyond me. Astrology, I have not mastered.
I cannot and will not live by predicting fathers' deaths.[10]

Galen did very well in Rome, the busy center of the Empire. Between his thirty-second and thirty-seventh years he entered the circle of those about the Roman court and successfully competed with other physicians and intellectuals for preferment, until, as he proudly relates, Marcus Aurelius called him "the first of physicians and the only philosopher" (K 14.660). During this period, I believe, Galen created a picture of himself and of Hippocrates that was intended to justify such compliments as the emperor's: in a slack and decadent world Galen upheld the highest ideals in medicine and philosophy and called his contemporaries back to the Classical as he found it in Hippocrates

---

9. See *On Sects for Beginners, Scr. Min.* 3.1–32, written early in his first Roman period (cf. *Scr. Min.* 2.93–94). *Subfiguratio Empirica,* another work on sects dating probably from the same time, outlines the logical position of the Empirics and opposes to it the dogmatism of Hippocrates as Galen then defined him (see Karl Deichgräber, *Die griechische Empirikerschule* [Berlin, 1930], pp. 42–90). *On the Best Sect* is of dubious authenticity, and I leave it out of my account.

10. Juvenal, *Satire* 3.41–44:
Quid Romae faciam? mentiri nescio; librum,
si malus est, nequeo laudare et poscere; motus
astrorum ignoro; funus promittere patris
nec volo nec possum.

and Plato. Nor was he shy about saying so. There is a large element of rhetorical commonplace and exaggeration in his descriptions of himself, but that does not necessarily reflect on their sincerity. A fascinating tale in itself, Galen's struggle in Rome is relevant to our considerations because of its effect on the image of Hippocrates and Hippocratic medicine which he created and because he rewrote the history of medicine according to the rhetorical posture he assumed.[11]

Two physicians in particular, who had arrived there before him, became the objects of Galen's enmity as he jostled for position: Antigenes, who "was considered to be the leading physician and who had everyone of high standing among his patients" (K 14.613), and Martialius, "who at the time had long had the reputation among young physicians of being a great anatomist; two of his books on anatomy were famous" (K 14.614–615).[12] Galen seems to have taken over the patients of the former and the reputation of the latter.

Martialius was an Erasistratean, "vicious and quarrelsome," says Galen, "despite being over seventy years of age" (*Scr. Min.* 2.94–95). When he heard about Galen's public displays of anatomical dissection, he asked Galen whom he most admired of

11. Galen gives a connected narrative of his first residence in Rome, his feuds with other physicians, and his rise to preferment in *Prognosis to Epigenes* (K 14.599–673). He gives further details elsewhere, particularly in his works on phlebotomy (K 11.147–249) and in his works on his own books (*Scr. Min.* 2.80–124). For the names of the various members of the aristocratic and imperial circle of Rome with which Galen associated, see the article on Galen in *Prosopographia Imperii Romani*, editio altera, ed. Edmund Groag and Arthur Stein (Berlin, 1933–1970), pt. 4, pp. 4–6. The article on Eudemus in the same work, pt. 3, p. 90, contains numerous errors which stem from confusion of the Empiric physician in Pergamum with the Peripatetic philosopher of the same name whom Galen knew in Rome. The errors may have come originally from Johannes Ilberg, "Aus Galens Praxis," *Neue Jahrbücher für klassische Altertum* 15 (1905), 286–287. See now also Vivien Nutton, "The Chronology of Galen's Early Career," *Classical Quarterly* 23 (1973), 159. Glen W. Bowersock, *Greek Sophists in the Roman Empire* (Oxford, 1969), pp. 89–100, gives a general description of public professional quarrels in Galen's time which is useful background for Galen's own experience.
12. The man who appears as Martialius in our texts of *On His Own Books* is the same as the "Martianus" of Kühn's text of *Prognosis to Epigenes*. Ilberg (1, 209–210) made the identification, but he did not explore its implications. Antigenes was a student of Quintus and of Marinus (K 14.613).

the "ancients," and praised Erasistratus as superb in anatomy and all else in medical science. Not one to forego a challenge, apparently, Galen responded by writing *Anatomy according to Hippocrates* in six books and *Anatomy according to Erasistratus* in three. His purpose was to prove that all the results of anatomical dissection were known to and used by Hippocrates, whereas Erasistratus' work was inferior and incomplete. The works have perished, but Galen's frequent mentions of his essay on Hippocratic anatomy give some notion of its contents.[13] It included osteology, drawn from the surgical works of the Hippocratic Corpus, the anatomy of the blood vessels and nervous system as it appears in the cryptic passage of *Epidemics* 2, and material on the system of digestion and nutrition. Galen writes that he spoke of the "shamelessness of the Empiric physicians who dared to call Hippocrates an Empiric" and that he argued that Hippocrates "pursued anatomy intensely, because it helped the science" (K 18 A.524).[14] His manner of attack on Erasistratean anatomy is less clear. I assume from similar quarrels to be described below that he attacked Martialius' books. He apparently gave a bold performance that rewrote the history of anatomy, and it is unfortunate that the works did not survive.

Galen's attacks on Erasistratus and Martialius did not stop at anatomy. He tells of a public debate in which he humiliated the Erasistrateans on the subject of phlebotomy; and he has left us the substance of the debate, *On Phlebotomy, Against Erasistratus* (K

13. The one fragment of the work that I know is given in Richard Walzer, *Galen on Jews and Christians* (London, 1949), p. 11. (Galen appears to have said that physicians who practice without scientific insight are like Jews following the laws of Moses, who do not present reasoning but say, "God spoke, God commanded.") Galen speaks of the work and its contents at the following places: K 1.481, 18A.524, *Scr. Min.* 3.111, and *Opinions of Hippocrates and Plato,* Müller 515, 574, 607, 647. Hunain Ibn Ishaq knew the work and said of it in his catalogue that Galen's purpose in it was to prove Hippocrates' familiarity with dissection, for which he drew evidence from all his works. Cf. Walzer, *Galen on Jews and Christians,* p. 18, Gotthelf Bergsträsser, "Neue Materialien zu Hunain Ibn Ishaq's Galenbibliographie," *Abhandlungen der Kunde des Morgenland* 19.2 (Leipzig, 1932), 27.
14. I would infer that Martialius spoke of Hippocrates as an Empiric who did not dissect, as opposed to his own "logical" Erasistratus. Galen disputes that correct historical view and shapes his argument to counter it.

11.147–186). The argument is a good example of what Galen's rhetoric does with medical history.

Galen begins his argument mildly, wondering why so knowledgeable and precise a physician as Erasistratus seems to have
had nothing to do with such an effective and honored remedy as
phlebotomy (K 11.147). He then quotes Erasistratus' one reference to phlebotomy, a passage in which he praised his teacher
Chrysippus for binding off the limbs of patients who were bringing up blood: this, said Erasistratus, accomplishes what
phlebotomy would, in reducing blood in the thorax, but it also
preserves the nourishment in the bound-off blood for the body's
use when danger is past (K 11.148–149).

In refuting Erasistratus, Galen argues first that no sensible
reasons for ignoring phlebotomy are given by Erasistratus nor
by other students of Chrysippus and that arguments that have
been offered about the dangers of phlebotomy are silly and
contradictory (K 11.151–152). Next Galen summarizes Erasistratus' theory that diseases involving inflammation without
wounds come from a plethora of nourishment in the veins. If
this is true, argues Galen, phlebotomy is a kinder and more
effective cure than the starvation Erasistratus used (K 11.153–
161). As he proceeds, Galen gradually becomes more abusive
about Erasistratus' stupidity or carelessness. "Maybe, as they say,
you had contempt for seeing the ill, and stayed at home to write
up your thoughts. But even so, you could have read Hippocrates"
(K 11.159). He cites scattered passages in Hippocratic works and
quotes the case of the servant of Stymargos, from *Epidemics* 2 (L
5.127), arguing that it illustrates clearly the reasoning behind
phlebotomy (K 11.160–162).

At this point Galen says, "But I don't want to pain the Erasistrateans overmuch by praising Hippocrates. They seem to me,
and so does Erasistratus before them, to want to quarrel with
Hippocrates. So let us leave the Hippocratic material." Galen
proceeds to list the dogmatic physicians who used phlebotomy:
Diocles, Pleistonicus, Dieuches, Mnesitheus, Praxagoras,
Phylotimus, Herophilus, Asclepiades, Mantias, Athenaeus,
Agathinus, Archigenes. And so did the whole chorus of Empirics, Galen adds.

Next Galen cites Nature's own bloodletting in menses, lochial flows, and milk, quoting Hippocrates that a woman will not get podagra if her menses are regular. "But why," adds Galen, "should I quote Hippocrates to a man who is his enemy? I think the truth herself will speak to you through my witness" (K 11.165).

After listing conditions cured by phlebotomy, Galen proceeds, "You are not quibbling with Hippocrates alone if you don't use the remedy, but with the experience of all physicians and with human life. . . . It is not only Hippocrates' view, but that of all men. Yet you seem to me, because of desire to win over Hippocrates, to have become more senseless than anyone" (K 11.166–167).

Having slipped in this manner into the accusation that Erasistratus' motives are enmity and jealousy of Hippocrates, Galen repeats the charges (K 11.168, 169) as he again names his list of authorities and describes the ways in which phlebotomy is efficacious and again ridicules Erasistratus for ignoring a remedy that so nearly fits his own view of the source of disease. For the remainder of the treatise (K 11.167–186), Galen gives a more detailed argument that phlebotomy is the safest and most efficacious way of reducing plethora in patients, especially those in danger of bringing up blood. The Erasistrateans, Galen argues (and he means Martialius), do not even understand Erasistratus and, of course, cannot defend him (K 11.194–195).

Galen's debating style projects onto history his own emotions and his situation: Erasistratus, like Martialius, was blinded by jealousy and unable to pursue truth. Any sensible person would have seen Hippocrates' superiority, and only enmity, ignorance, or laziness could account for failure to do so, Galen infers. Galen's rhetorical device has had serious results. While, in fact, Erasistratus did not mention Hippocrates and showed no interest in him, Galen's debate required otherwise. Galen held the public disputation at age thirty-four. Looking back some thirty years later, he seems embarrassed at his contentious rhetoric. He says that when he returned to Rome the second time (at age forty), he swore to give no more such public disputations because his medical practice had been even more successful than he had hoped and he knew that such things only caused enmity (*Scr. Min.* 2.94–96). But in a work actually written near the age of forty he chortles over the success of his debate: while Erasistrateans had previously murdered patients by refusing to bleed them, now they had been overwhelmed by his arguments and were bleeding indiscriminately and in Erasistratus' name. He will therefore lay out the science precisely, after proving that in fact Erasistratus had rejected phlebotomy (K 11.187–249).

Galen claims that he made his medical reputation most dramatically by making predictions that were fulfilled and by disputing his competitors' diagnoses and manner of treatment. Typically, Galen reports, his competitors would gossip behind his back, ridicule his predictions, and hope for the worst. Then when he succeeded, they would be abashed and humiliated. Galen would insist modestly that his science was all written in Hippocrates' works of which they were ignorant. There was no magic to it (K 14.651, 673). A crucial success which Galen describes at length was his prediction of the course of the quartan fever of Eudemus, a Peripatetic philosopher and teacher, who was popular in the aristocratic circle. Eudemus was sixty-three years old. Galen's competitors predicted the worst, but Galen took over the case and accurately predicted the periods of the fever and the time of the successful crisis. Both Antigenes and Martialius were humbled to the dust by the results and showed their hatred (K 14.605–620). Galen describes his conversation with the newly healthy philosopher, in which they talked of the depravity of the society and the age. Eudemus warned Galen that his competitors were greedy and dangerous men who came to Rome to make their fortune by practicing wickedness as much as medicine. They were the sort of men who use geometry and arithmetic only to calculate their expenses and who study the stars and prediction only to find out whose money they will inherit (K 14.604–605. For the rhetorical commonplace, cf. the quote from Juvenal above.). If Galen was not careful, they would poison him, as one young, excellent physician had been poisoned ten years earlier. (The great Quintus, Galen observes, had been driven from Rome on the charge of killing patients, K 14.602.) Galen and Eudemus agreed that Galen, because of his

private means and superior background, could hardly be understood or appreciated by such men. Galen decided to leave Rome
as soon as he could. In the meantime, he would protect himself
(K 14.620–624).

Galen's work *The Best Physician Is Also a Philosopher* explores
the same theme. It describes Hippocrates, the ideal physician,
who deserves emulation but does not receive it from the ignorant, grasping, and lazy physicians of Galen's day. A man cannot
care for science and money both, says Galen. If he follows Hippocrates he will limit his concern for wealth to avoiding hunger,
thirst, and cold. He will pay no attention to Artaxerxes and
Perdiccas (that is, royalty and its retinue). He will heal the poor
in Kranon, Thasos, and other small towns. He will leave his
students to care for the people in Cos, while he himself will
wander everywhere to observe all Hellas. To judge by observation what is taught in theory, he must observe cities and their
environments. The true physician is a friend to self-restraint and
a companion to truth. He must train himself in logical method
and know the genera and species of disease and how to find
indications for treatment of each. He must study the body and
its elemental structure, the usefulness and function of each part,
and so forth (*Scr. Min.* 2.5–6). The work is an appropriate assertion for the young Galen, who was then planning to leave
Rome.[15] It also reflects the subject matter of other of his writings
from the same period, as we shall notice. Galen adds to his list of
the sins of his contemporaries their bad literary style: Hippocrates "was concerned with style, but they are so far the opposite
that one can see many of them making two errors in a single
word, difficult as that is to imagine" (*Scr. Min.* 2.2).

Flavius Boethus, a Roman of consular rank with connections at
the court, became Galen's close friend and sponsor. Boethus was

15. The close relation of *The Best Physician Is Also a Philosopher* to Galen's immediate concerns in his early stay in Rome and its failure to cite earlier
works lead me to date it confidently to this point in his career. Ilberg and
Bardong do not seem to treat the question. Galen himself does not cite the
work earlier than his catalogue of his own books. Hans Diller, "Zur Hippokratesauffassung des Galenos," *Hermes* 68 (1933), 176–181, remarked its odd
manner of citing *Places in Man* and placed it late in Galen's career. I would
urge that early dating better solves the difficulty.

interested in Aristotelian philosophy and in anatomy (Galen's works on Hippocratean and Erasistratean anatomy were dedicated to him). Galen describes another of his great medical and social successes in curing Boethus' wife, who was suffering from a mysterious and persistent "female flux" from the genitals. The story is interesting enough to deserve retelling. She was in the direct care of midwives, who took their directions from the male physicians, many of whom, including Galen, consulted together about the case. The flux persisted despite drying regimen and styptic ointments, and the lady swelled up as though pregnant. One day at the bath she had an intense pain, like labor, and passed watery fluid. She was carried from the bath in a faint, just as Galen happened to be passing by. He advised and assisted her attendants. He rubbed her extremities and her stomach to restore her. That night he had a kind of waking vision about her condition, remembering how she had felt under the rubbing: like watery cheese, trying to turn to cheese, but not yet cheese.[16] From that he analyzed the past errors of her physicians and worked out the rationale of treatment: give her the least possible drink, much massage, diuretics, purges, and boiled honey for ointment, all to evacuate the moisture in all possible ways. He proposed to Boethus that he, Galen, take over the case privately. Boethus agreed, and on the success of the treatment within a month, gave Galen four hundred gold pieces and praised him to the other physicians and to the court, much to his competitors' discomfort (K 14.641–647).

Galen's association with Boethus and the Peripatetic philosophical circle was the source of other writings and other elaborations of his view of Hippocrates. His expertise in dissection and his own anatomical discoveries, along with his extensive training in philosophy, made him peculiarly capable of contributing to discussion of the important philosophical questions of the day: how is the universe constructed and does it have a

16. The analogy between cottage cheese (or perhaps yogurt) and what is left over from food after the nutritive elements are removed appears elsewhere in Galen's works (cf. K 2.614, CMG 5.4.2, p. 31). *Perittomata* of the body's nutrition, which are like curds and whey, are excreted as phlegm and sweat.

purpose, how should we describe the nature of the human soul and what is its relation to the body, what makes the body work and where is its controlling factor. On the last question, Galen's discovery of the recurrent laryngeal nerve gave the opportunity for a dramatic contribution to argument. Aristotle and the Stoics had placed the reason and the controlling factor in the heart, Plato in the head. The discovery and mapping of the nerves by the Alexandrian anatomists had made the head more likely, but an argument in favor of the heart could still be made because the voice, the instrument of reason, issued from the chest. Galen proved that the voice is controlled from the brain, through the recurrent (running back) laryngeal nerves, which issue from the spinal cord in bundles of nerves, arrive in those bundles at the chest, then reascend to the larynx, right and left, where they control the opening and closing of cartilaginous plates that are struck by issuing air.[17] Galen apparently had also refined medical knowledge of the mechanism of breathing by adding the description of the function of the intercostal muscles to his teacher Pelops' description of the function of the diaphragm (K 18 B.927). Galen could, therefore, in public anatomical display, using live animals, prove how the mechanism of breathing and voice production is controlled and whence, by cutting, one by one, the nerves involved. With one stroke (quite a stroke!) he could prove that the brain controls the voice, by severing the recurrent laryngeal nerve and impairing no other faculty. The demonstration was most convenient on large-voiced animals, pigs and goats (K 14.627), but dramatic on long-necked birds in whom the nerve had to make an immense circuit.

17. Galen describes his discovery of the recurrent laryngeal nerve in *On the Usefulness of the Parts: De usu partium,* ed. Georg Helmreich (Leipzig, 1907–1909), I, 412–425; cf. the translation and notes of Margaret Tallmadge May, *Galen On the Usefulness of the Parts of the Body* (Ithaca, N.Y., 1968), I, 361–371. Joseph Walsh, "Galen's Discovery and Promulgation of the Function of the Recurrent Laryngeal Nerve," *Annals of Medical History* 8 (1926), 176–184, gives a pleasant account of the discovery and its significance. Walsh dates the discovery to Galen's Pergamene practice, but gives no evidence, and I find none. Galen dates the demonstration itself to fifteen years before ca. 177, which would place it very early in his first stay in Rome (K 14.628–630). Walsh may be correct that the discovery was made earlier, in Pergamum.

Boethus arranged the demonstration, procured the animals, and assembled Stoics and Peripatetics to be instructed (K 14.626). Despite the boorishness of the eminent Peripatetic Alexander of Damascus, who threatened that he would not believe the evidence of his senses, the display went forward, and Galen writes that he convinced everyone. Boethus provided shorthand recorders to whom Galen dictated what had been shown, thus producing, apparently, his work on *Voice and Breathing,* which has been lost (K 14.627–730). Plato was right about the tripartite nature of the soul and the seat of the reason, Aristotle and others wrong, Galen proved.

Galen extended his contribution to philosophical debates by arguing that Plato was not only, on the whole, right in philosophical, logical, and moral questions, but furthermore he was the follower of Hippocrates, from whom he got his main doctrines! Where Plato was wrong or questionable, Hippocrates was right or had no view because the topic is irrelevant to medicine. Galen's extended argument on the subject was offered in the *Opinions of Hippocrates and Plato,* dedicated to Boethus (of the work's nine books, the first six were written during Galen's first residence in Rome). Galen argued in *On the Usefulness of the Parts of the Body* (of the seventeen books, only the first was written during Galen's first stay in Rome, and dedicated to Boethus) that Aristotle's teleological treatment of the parts of the organism is derived basically from Hippocrates! Furthermore, Galen argued in *Elements according to Hippocrates* that the Aristotelian elemental theory is already all there in Hippocrates and that Plato and Aristotle were simply following the Master in the subject![18]

18. The work on *Elements according to Hippocrates* seems to me to represent efforts at composition from two different periods of Galen's career, as do *Usefulness of the Parts* and *Opinions of Hippocrates and Plato.* Ilberg (2, 504–505) comments on the queer structure of the work, but dates it next to *Temperaments,* which is mentioned in the final portion. Galen originally wrote the work as a going away present for a friend (unnamed, CMG 5.9.1., p. 3). The original work consists of the continuous argument about elements (K 1.413–480). After that, Galen says, "Some people want me to write a second *logos.* Now I will put a head on this one" (K 1.481). The second part seems to me to be a continuation written early in his second residence in

Galen develops an interesting shifting approach to argument in his early argumentative works which were written during this first stay in Rome. Eclecticism, which involves the effacement or reconciliation of differences in approach and attitude (as between Plato and Aristotle), is important to him, but equally important is rhetorical exaggeration of differences when the argument goes that direction. Galen conjures up opponents or potential opponents in virtually every work, with the result that he can shift, when his argument is weak, from exposition of his medical subject to attack on his opponents' insight, morals, verbal style, or attitude about the past, and particularly to attack on their attitude about Hippocrates. Galen develops the habit of padding out exposition of truth with the expression of righteous indignation at the stupidities of those who have not seen it. He developed an interpretation of history according to which everyone after Hippocrates who had been wrong was perversely and stupidly wrong because he refused to see the truth as Hippocrates had seen it. He generally attacks, not his contemporaries, but the prominent figures from the past whom they admired, and in the process he readjusts the history of medicine by projecting his own quarrels into the past. I shall take note here of the substance and the manner of argument of the relevant works as they affect Hippocrates and medical history, since they are not well known and have not been studied from this point of view.

Elements according to Hippocrates asserts that Hippocrates is *archegete* (first founder) in elemental philosophy as well as in medicine. Galen admits that he is the first who ever realized that fact and tells how he arrived at it. At the age of nineteen he was instructed in the opinions of the first-century A.D. pneumatic theorist Athenaeus of Attalia that hot, cold, wet, and dry are the obvious elemental components of the body, as of all matter. But

---

Rome, when he was working out more fully the aspects of his system, which included *Temperaments* and the works on drugs, written in fact between those two works, the former of which he cites and the latter of which he promises (K 1.490). The added portion responds to objections that Hippocrates' *Nature of Man* is not about elements in either title or substance, and it offers a transition to *Temperaments* and the works on drugs.

Galen saw in Athenaeus' account an ambiguity between "the hot" (or "hotness") and "the hot thing" (both *to thermon* in Greek), that is, between *fire*, the visible element which is the perfectly hot thing, and the quality *hot* which characterizes things that are more hot than cold. He challenged his teacher, who could not understand the problem and called Galen a quibbling sophist. Galen held his peace, but pondered the problem (and, like his father, mused on the necessity for logical theory). He finally realized, so he tells us, that fire, water, earth, and air must be the obvious, that is, visible, primal elements, and that they are characterized by the qualities hot, wet, cold, and dry which inform the primal, undifferentiated matter. Galen had, of course, reproduced Aristotle's reasoning, and he attributed this line of reasoning to Hippocrates, reading it into the medical work *Nature of Man*. That is the thesis of *Elements according to Hippocrates*. As Galen puts it:

> Hippocrates not only says as he proceeds in the work *Nature of Man* that these are the elements, but he indicates their qualities by which they naturally affect and are affected. And he was the very first to define them. But most people, since they do not understand him because of the ambiguity of his account, are confused, like Athenaeus of Attalia.... Virtually none of more recent physicians has laid out an account of the whole of medicine as Athaeneus did. Nevertheless he is demonstrably in error in this as in many other matters. And so is everyone else. I do not know anyone who took in hand the ancient medicine and perfected the methods the ancients bequeathed to us. In fact, if one must tell the truth, they have removed much that was correct, as does Athenaeus when he says that the elements are obvious and need no demonstration.   [K 1.456-457]

During his second residence in Rome, when Galen was writing his more mature and comprehensive medical works, he extended this book on elements to include discussion of secondary elements of the body, the four humors, which in turn comprise the primary bodily elements that in turn make up the organic parts, thus integrating the elemental theory with his comprehensive medical system. Finally, late in his career (about twenty-five years later), he wrote a commentary on the *Nature of Man* in which he recalled and defended his early assertions about Hippocrates' philosophy of medicine.

In his work *On the Opinions of Hippocrates and Plato,* written in its early form during his first residence in Rome, Galen argued that the philosophical views of Hippocrates and Plato are the same, if one allows for their differences of approach, and that they are correct.[19] When he wrote the continuation, books 7–9, Galen said that the first six books, which he had dedicated to Boethus, had seriously shaken the Stoics, Peripatetics, and physicians at whom he had directed them. Some, he said, have already publicly changed their positions regarding the source of the nerves and the seat of the governing part of the soul (pp. 582–583. Compare his crowing about the Erasistrateans and phlebotomy.). Grand as is his design of proving that Hippocrates and Plato were essentially one and correct in philosophy, Galen only attempted to show that they agreed on "the powers that govern us, their number, the nature of each, and the place that each occupies in the body" when he wrote the first six books. From those, "nearly all particular details follow" (p. 168). How he demonstrated that much is in part unclear because the part of book 1 that discussed the nervous system is missing, but he seems to have reused the material from the *Anatomy according to Hippocrates.* In the extant part he quotes at length the passage from *Epidemics* 2 which shows that Hippocrates knew the anatomy of veins and arteries (pp. 471–474) and repeatedly quotes the statement from *Nutriment* that the "rooting of veins is the liver, of arteries the heart; from there blood and *pneuma* travel to all parts, and heat goes through them." That Plato believed in a tripartite soul, resident in head, heart, and liver, is easy to prove. With effort, Hippocrates' anticipation of the doctrine can be inferred. Plato, says Galen, wrote about the powers of the soul, whereas Hippocrates wrote about bodily organs. Effectively the two divided the subject (p. 471). That Plato fol-

19. I cite *On the Opinions of Hippocrates and Plato* according to the pagination of the edition of Iwan von Müller. I am very grateful for having seen the text and translation prepared by Phillip De Lacy for the Corpus Medicorum Graecorum.

lowed Hippocrates can be inferred from Plato's use of Hippocrates in the *Phaedrus,* where he credits his own method of division to Hippocrates.[20]

Galen devotes the bulk of books 1–6 to refutation and ridicule of people who are too stupid and jealous to agree with Hippocrates and Plato, among them the physician Praxagoras of Cos, who, "since he did not see any nerves growing out of the heart, but yet was ambitious to compete with Hippocrates at all costs, and wished at all costs to eliminate the brain as the source of the nerves, ventured on no inconsiderable fiction," namely, that the arteries get smaller and smaller and turn into nerves (p. 144. Similarly, he later says that Erasistratus was competing with Cos rather than honoring the truth [pp. 688–689].). This polemical stance represents an advance over Galen's posture in his youthful work *On the Anatomy of the Uterus,* in which he described Praxagoras, like the other early physicians, as an honorable man who correctly explained observations which also occur in Hippocrates. The Stoic Chrysippus receives the largest amount of refutation and ridicule because, Galen says (p. 583), after he had refuted him in book 2 someone said that the refutation was insufficient and so he set out to devastate him in the three following books. The result is a rather shapeless treatise that he presented to Boethus.

Books 7–9, added about ten years later, are more positive in their argument, though their use of Hippocrates is still somewhat elusive. For his proofs about the nature of Hippocrates' elemental theory, Galen refers the reader to his work *On Elements according to Hippocrates.* In his discussion of logical theory in book 9, however, he breaks new ground in his Hippocratic interpretation. He cites from Plato discussions of the way in which one distinguishes between things and concepts that are alike in some respects. It is clear that Plato discusses such subjects with an eye on logical theory. To show Hippocrates' primacy, Galen quotes from the Hippocratic Corpus statements about making

20. For a summary of Galen's treatment of Plato and Platonic philosophy, see Phillip De Lacy, "Galen's Platonism," *American Journal of Philology* 93 (1972), 27–39.

distinctions in diagnosis of diseases, in estimating the condition of patients, and in judging nutritional virtues of different wines.[21] Hippocrates' works do not deal explicitly in logical methodology, but they do illustrate the vocabulary with which one describes similarity and difference. Galen's crucial passage (one which his teacher Aephicianus had used to prove Hippocrates a Stoic) is the opening passage from *The Surgery*, which I translate as follows: "Things like or unlike, beginning from the largest, the easiest, the completely understood, which can be seen, touched, heard. Things that can be perceived by sight, touch, hearing, the nose, the tongue, and the intelligence (*gnômê*). Things knowable in all the ways that we know" (L 3.272). *The Surgery* goes on to describe the requisites for operations: the patient, the operator, the assistants, the instruments, the light. Galen puts beside the above passage statements from *Joints* and *Prognostic* urging that one should compare the patient's current condition with his normal appearance and a passage from *Regimen in Acute Diseases* on the different effects of different wines. From those he proves that Hippocrates had a logical terminology: *skeptesthai*, investigate; *sêmeia*, indications; *dihorizesthai*, distinguish. The fact that Hippocrates used the word *gnômê* in *The Surgery*, Galen argues, proves that he gave reason equal status with sense perception and that he avoided the excesses both of empiricism and of dogmatism. Whatever we think of Galen's standards of proof, we must admire the boldness of his argument. For our analysis of the Hippocratic tradition it is significant that Galen is original and that he is conquering new territory for Hippocrates. He builds on hints by predecessors and uses his teachers' methods of Hippocratic interpretation, but the results are new syntheses that alter historical relationships.

21. Galen quotes from the surgical works and from *Prognostic* and *Regimen in Acute Diseases*, "most genuine works," on which he began commentaries probably soon after he finished this continuation of *On the Opinions of Hippocrates and Plato*. Galen's discussion of *Regimen in Acute Diseases*, which says repeatedly that Hippocrates is confused and expresses himself badly, seems to have been added still later, after Galen had written his commentary on *Regimen in Acute Diseases*.

On the Usefulness of the Parts of the human body, book 1 of which was written for Boethus, argues that the parts of man, the intelligent and godlike being, were all formed perfectly for the use for which they were intended.[22] Book 1 lays out the thesis and illustrates it by describing in detail the construction of the hand, that magnificent instrument so characteristic of the intelligent being. Galen says that his work is necessary because his predecessors, even Aristotle and Herophilus, are deficient on the subject and do not understand Hippocrates' writings. But not even his writings are adequate, "since he treats some subjects obscurely and omits others altogether, though in my estimation, at any rate, he has written nothing that is incorrect. For all these reasons, then, I have felt moved to write a complete account of the usefulness of each of the parts, and I shall accordingly interpret those observations of Hippocrates which are too obscure and add others of my own, arrived at by the method he has handed down to us" (1.15, trans. May).

As Hippocrates' basic philosophical statement, Galen cites one of the cryptic aphorisms from *Nutriment:* "Considered as a whole, all parts in sympathy; considering the parts of each part, they are for a function (*ergon*)."[23] Galen explains:

> No one is ignorant what the function of the hands is. It is obvious that they are for grasping. But that all their parts are of a kind and size for cooperating in a single function not everyone understands. But Hippocrates did understand it that way, and I propose now to demonstrate that very thing. On the basis of it is constructed the method of discovering the usefulness (of the parts), and the errors of those who hold views contrary to truth are refuted.   [1.13–14]

22. I cite *On the Usefulness of the Parts* according to the pagination of Georg Helmreich, 2 vols. (Leipzig, 1907–1909), by volume and page. The work is conveniently available in the English translation by Margaret May.

23. I have attempted to translate the reading of the Mss. as indicated in Helmreich's apparatus (1.12). Helmreich, considering the vulgate text of *Nutriment,* alters the reading to "considered part by part the parts in each part aim at a function." May translates Helmreich's text: "Taken as a whole, all the parts in sympathy, but taken severally, the parts in each part cooperate for its effect." The text is problematic, but the textual problems are not crucial to Galen's point, nor to mine.

Galen's first example is the fingernails: Plato, "though a follower (*zelôtês*) of Hippocrates if anyone was, having received his most important beliefs from him," nevertheless made fun of the fingernails as useless: rudimentary claws. Aristotle said that they were for protection, but he did not say from what. No one has pointed out how useful they are for picking up small, hard objects and many other purposes. But Hippocrates clearly knew all this because he told us how long they should be: "The nails neither to project beyond, nor fall short of, the finger tips." And clearly they can fulfill all their functions best if they are proportioned as Hippocrates prescribes (1.10–12).

Galen offers an interpretation of one other passage from Hippocrates in book 1, after which he eschews further demonstration of Hippocrates' understanding of the subject. The second passage is also from chapter 4 of *The Surgery*. The context of the quotation is a description of the necessary instruments and arrangements for the surgical office and advice to the surgeon how to sit and stand, where to place the instruments, and so forth. Here is chapter 4 as translated by E. T. Withington, with the part quoted by Galen in italics:

> The nails neither to exceed nor come short of the finger tips. Practise using the finger ends especially with the forefinger opposed to the thumb with the whole hand held palm downwards and both hands opposed. *Good formation of the fingers: one with wide intervals and with the thumb opposed to the forefinger,* but there is obviously a harmful disorder in those who, either congenitally or through nurture, habitually hold down the thumb under the fingers. Practise all the operations, performing them with each hand and with both together—for they are both alike—your object being to attain ability, grace, speed, painlessness, elegance, and readiness.

Galen quotes only the italicized passage (having previously quoted the sentence on the fingernails), and he comments, as translated by Margaret May:

> In fact, the division took place for the sake of enabling the fingers to spread apart to the greatest extent, often a very useful position.

And so he properly says that it is particularly when the fingers have that [ability] for the sake of which they were formed that their construction is most advantageous. For surely, to this construction is due also the opposition of the thumb to the other fingers, since if the hand were merely divided into fingers and the thumb not set farthest from the others, it would not be opposable to them. Verily, here too Hippocrates teaches many things in but few words to those, at least, able to understand what he says. Hence, when I have once called attention to the method of exposition found in all his writings, it will perhaps be proper for me to imitate not only the other virtues of the man but also this very trait in him of teaching much in few words and to abstain from going over all his sayings in detail. Except in passing, therefore, I do not propose [in every instance] to state that Hippocrates had an excellent understanding of such matters; my purpose is rather to discuss in detail the usefulness of all the parts.   [1,16–17]

In the final sixteen books, in which he carries through his argument for design in magnificent fashion, Galen makes no attempt to prove that Hippocrates held the views he is presenting, nor does he again assert that his method of investigation is derived from Hippocrates. He does, however, offer a sprinkling of quotations from and references to Hippocrates in his usual manner: Hippocrates used to say Nature was just (1.116, 2.50, 116, 376); Hippocrates was right to say blows on the temple are serious. Even before Hippocrates, Nature realized that she needed to protect the temporal muscles (2.118); sophists who deny the providence of Nature mock Hippocrates, who thought we should imitate what Nature does in crises (2.451); and so on. Galen's failure to attempt further proof or to reiterate his dependence on Hippocratic method does not mean, however, that he had abandoned the claim, as we shall see from his descriptions of his method in *Hygiene* and *Abnormal Breathing*, which were written in the same period as the last sixteen books of the *Usefulness of the Parts*.

If *On the Best Sect* is by Galen, it would appear to have been written in this same period, as also was a lost work on *The Plague according to Thucydides*. The latter contrasted the layman's superficial description of a disease, which Thucydides exemplifies, with the professional selection of significant phenomena exemplified in Hippocrates' *Epidemics* (K 7.850–851). *On the Best Sect* defines science as those specialist matters unavailable to laymen. It argues that the logical methods claimed by Empirics and Methodists do not in fact work and that both sects actually reason from hidden causes when they practice medicine, whether or not they admit it (K 1.142–155). Hence, it argues, the *logos* is indispensable and the logical sect is best. Hippocrates is cited in this work only to bolster the arguments of the logical sect (see, for example, pp. 184, 197, 201, 208, 213) in contrast to use of him in the work on *Medical Experience.* I am not satisfied that this is Galen's work. Ilberg (4, 603–605), Bardong (612–614, 633), and Iwan von Müller[24] have disagreed about whether it is genuine and, if so, when it was written. Galen speaks of having written a work entitled *On the Best Sect,* which he classed among his logical works, not an unlikely description of the extant work (*Scr. Min.* 2.120). Nonetheless, I incline to think that this is not Galen's work, largely because of the way in which it uses Dexippus and Apollonius (K 1.144). They exemplify the problem, "how do we set criteria," very much as Erasistratus used them. Galen is nowhere so noncommittal about the correctness of Hippocrates' students' views.

In *Prognosis to Epigenes,* Galen describes how, after great and continuing success, he was about to be introduced to the emperor. But he carried out his decision to leave Rome, leaving secretly like a runaway slave to go home to Pergamum. Incidental to the occasion was a plague in Rome. There has been considerable controversy whether or not Galen left Rome to flee the plague. Recently there has been little interest in the subject, nor can I feel strongly about it. But at times it seemed an immediate question to those who would evaluate Galen. The most recent one I know was Charles Daremberg, who attempted in the mid-

24. "Ueber die dem Galenos zugeschreibene Abhandlung *Peri tes aristes haireseos,*" *Sitzungsberichte der Münchner Akademie der Wissenschaften,* philosophisch-historische Klasse (1898), 53–162.

nineteenth century to do for Galen what Littré had done for Hippocrates: physicians were very popular and influential at that time in Paris because of their heroic self-sacrifice during the cholera epidemic of 1832. To have fled the city was tantamount to disgrace.[25]

Whatever the reasons for Galen's flight, he was summoned by Marcus Aurelius three years later and apparently had to come. Yet he was permitted to stay behind in Rome while Marcus went to the German frontier for war with the Marcomanni. He retired to the countryside to be near the emperor's heir Commodus and to avoid the argumentative atmosphere of Rome. In the years of grace so achieved he finished works he had begun earlier and entered a new stage of his literary career in which he produced the writings that present his mature medical system (K 14.648–651).

To summarize the conclusions so far: Galen inherited Hippocratism of different sorts, which he developed in a rhetorical manner in response to the competitive society of Rome. In answer to real and imagined slights of himself or Hippocrates, Galen magnified his claims about Hippocrates, apparently beyond any that had been made by his predecessors, particularly claims about anatomical knowledge and philosophical coherence in the Hippocratic Corpus. At the same time, I detect in Galen a development away from an early tolerance for Empiricism, and from claims that Hippocrates was empiric in outlook, toward confirmed dogmatism and insistence that Hippocrates had the same outlook. I see no indication that Galen had to defend his interpretation of Hippocrates against others who practiced medicine in Hippocrates' name.[26] The Erasistrateans (specifically Martialius) and the Methodists appear to be the competitors against whom Galen created his unhistorical history of medicine.

25. Cf. Erwin H. Ackerknecht, *Medicine at the Paris Hospital, 1794–1848* (Baltimore, 1961), p. 185; Joseph Walsh, "Refutation of the Charges of Cowardice Made against Galen," *Annals of Medical History* n.s., 3 (1931), 195–208.
26. Galen does say that Martialius referred him to *Prorrhetic* 2, which describes physicians who make flamboyant predictions in order to attract attention (K 14.620).

GALEN'S SECOND RESIDENCE IN ROME

The six years after Galen's return to Italy, 169–175 A.D., his fortieth to forty-sixth years, were a time of prodigious literary production. He wrote, or began, the major works of his medical system, those works, in fact, that make it appropriate to speak of his "medical system," as well as philosophical works.[27] In about 175 A.D., he could say that he had already dealt systematically with all medical science (CMG 5.10.2.1, p. 78). At that time he commenced his series of Hippocratic commentaries in connection with which he learned much that he had previously not known about the history of the Hippocratic text and its interpretation.

In dealing with the Hippocratism of the bulk of his medical work, it would be of little use to record all Galen's quotations of Hippocratic commonplaces, such as "do good, not harm," or all his lists of dogmatic physicians (headed by Hippocrates), who agree on a point, or all accusations that an opponent has either ignored or misunderstood Hippocrates or was his enemy. I shall notice such matters either where people have mistakenly made medical history out of Galen's rhetorical gestures or where Galen's polemics seem to offer insight into his predecessors' use of Hippocrates and into his reaction to it. Primarily, however, I shall try to characterize, at least by examples, the mass of technical material and the relation of Galen's Hippocratism to it. I am concerned whether, or to what extent, use of Hippocrates is simply a rhetorically applied antique patina on a strictly Hellenistic science and to what extent Galen's use of Hippocrates is traditional and how it is unlike that of his predecessors.

In his mature work Galen consolidates his point of view and extends his range in his writing to cover whatever is of current concern in medicine, attempting, as he says, to cover the whole field and, it appears, attempting to replace or to supersede his

27. I estimate that, from 169 to 175, Galen wrote, on the average, something over three pages daily, measured in pages in the Kühn edition, which are fairly small. This calculation includes estimates of some lost works of which we know the number of books. I get approximately two and a half pages daily for works that are extant.

predecessors. Galen generally describes himself as writing for
friends, and he wavers, in descriptions of himself, between excesses of modesty and of pride. We cannot think of his writings
as "coming out" and being advertised and distributed widely
year by year. In fact, we cannot think of the situation in any
terms applicable since the invention of printing. Nor can we
imagine that there were only a few copies of each work circulat-
ing among a coterie. How wide his audience, how broadcast his
publication, is a mystery. Galen apparently did not employ a
publisher (that is, a group of copyists making manuscripts to be
sold). But his works were widely circulated and sold during his
lifetime, whether sponsored by loving friends or done by book-
sellers. Forged works were sold under his name, as he tells us
(*On His Own Books, Scr. Min.* 2.91–92), justifying the very useful
account of his genuine works that he wrote in late years. Despite
his formal disclaimers, it seems clear that he did aim at wide
influence. Protestations of modesty, of never signing his name to
his own works, and the like, appear periodically and should be
taken as formalities, since equally often one finds him facing
vicious detractors in the battle for men's minds and insisting that
his readers refer to his earlier works. Some scholarly readers of
Galen have taken his formal protestations in earnest. There was,
of course, a genre of modest dedications, in Galen's time, which
he follows in his works. But we can take the word of Athenaeus,
Galen's urbane contemporary (who portrays Galen at his ban-
quet of wise men), that Galen astounded the world by the volume of his works, and by their range.[28] Galen could not have
expected that his success would be so great and that he would
become the standard of medicine for twelve or fourteen centuries, but he did try to become known. He was, therefore, attempting to impress on the world at large his version of Hippocratic science, although he maintains the explicit convention that
he is writing for a coterie of friends and, although he sometimes
mentions critics, he does not identify and answer them directly.

As I have suggested, Galen took the *Elements according to Hip-*

28. Athenaeus, *Deipnosophistae*, ed. and trans. Charles B. Gulick, Loeb Classical
Library (1927), 1.1.

pocrates, which he had written for one purpose some years before, and adapted it to serve as the philosophical prolegomenon
of his system generally. To its arguments about elements he
added an account of the four humors, blood, yellow bile, black
bile, and phlegm, which he considered mixtures of elemental
fire, air, water, and earth, mixtures whose relationships are also
ascertainable in the foods and medicines and environmental influences that affect the body in health and disease. He completed this theoretical structure with *Temperaments,* that is, mixtures (*De temperamentis* K 1.509–694), an account of the basic
organic structures of the body composed from mixtures of the
humors, each organ, and each of the kinds of tissue (*homoiomerê*)
that compose organs having a mixture or temperament peculiar
to itself, while each individual has his own normal temperament.
On this theory medicine must be based. In *On the Temperament
and Force of Simple Drugs* (K 11.379–12.377), he applies the
theory to the basic pharmacopea of his time. Other works on
compounding of drugs supplement it. Specific applications of
the theory are given in *The Best Constitution of the Body* (K 4.737–
749), *Marasmus* (wasting away from dryness, K 7.666–704),
Faculties of Foods, (CMG 5.4.2), *Anomalous Dyscrasies* (unusual dis-
turbances of balance in organs, a lost work), and others.

Galen's theory of temperaments, composition of the body,
and effects of foods and drugs has its roots far back in the Greek
view of how the body works and of the relation of environment
to health and disease. His peculiar version seems to come most
directly from the pneumatic school of medicine, which, under
Stoic influence, apparently worked out the four-element, four-
humor theory of disease and health and worked also on equiva-
lence of seasons to temperaments and on the temperaments of
foods and drugs more or less as Galen took up the subject.[29]
Galen disputes some of their details, but credits them with de-
veloping the science. For example, Galen disputes their limita-
tion of the temperaments to four and their simple equation of
temperaments to seasons. He adopts their terminology, *eucrasia*

29. Still basic to the study of the pneumatic school of medicine is Max
Wellmann's *Die Pneumatische Schule bis auf Archigines* (Berlin, 1895).

for good balance of humors (and of hot, cold, wet, and dry), *dyscrasia* for the unhealthy imbalance; but he insists that the *eucratos physis*, the healthy temperament, is a fifth basic condition, distinct from hot and wet mixtures (K 1.509–534). Other disputes and "corrections" of Pneumatics followed from that initial one.

Galen managed to write (actually, to dictate) such prodigious quantities because he used the basic material of other people's writings, adding what he thought appropriate. Thus, the work *Temperaments* outlines the subject as "the followers of Athenaeus" presented it (K 1.509–523; I think he used an outline of the subject by Athenaeus himself). But then Galen adds that although the Pneumatics cite Aristotle and Theophrastus, they do not really understand them, and there should be five, not four, basic temperaments. The next hundred pages consist of often rambling and repetitive argument on the point. Hippocrates figures in the argument only as one whose aphorisms about spring and whose descriptions of a hot and wet season followed by disease the Pneumatics should have heeded (K 1.527–530). As usual when he is using material from the Pneumatics, Galen is complimentary to his source, and only midly abusive, if at all. The Pneumatics were great organizers and chart makers, who apparently offered exhaustiveness of description as one test of their science. In this, too, Galen follows and improves upon them.

In the book on drugs, Galen excerpted recipes from his Hellenistic predecessors—a proper and traditional method which he acknowledges. But he adds his own experience of testing and inventing compounds, and again, rarely, he cites Hippocrates.[30]

Galen also wrote at length on the nature of the pulse, its

30. Cajus Fabricius' very useful book, *Galens Exzerpte aus älteren Pharmakologen* (Berlin, 1972), lists Hippocratic citations. Fabricius identifies excerpts from the people who are Galen's primary sources, of whom seven account for 90 percent of the excerpts. Almost all of the sources date from the first and early second centuries A.D. Max Neuburger's *History of Medicine*, trans. Earnest Playfair (London, 1910), is particularly useful for its exposition of Galen's pharmacological material. Georg Harig, *Bestimmung der Intensität im medizinischen System Galens* (Berlin, 1974), has offered a useful analysis of the pharmacological works.

causes, its usefulness in diagnosis and prognosis, and kinds of pulses.[31] Here, too, his antecedents are the Pneumatics, particularly Archigenes, whose work Galen "corrects" and "improves." Galen does not claim that one can learn about sphygmology from the Hippocratic Corpus. Hippocrates appears not to have been ignorant of the science of the pulse, but he did not work it out fully or develop the terminology (K 8.497). Galen considers his own writings on the pulse to be an improvement in consistency, accuracy, and thoroughness of classification of phenomena. His charts (for example, twenty-one kinds of pulses, twenty-seven anomalous pulses [K 8.504–506]) are intended to exhaust the possible combinations, such as the pulse that is long, broad, high, and large, and the anomalous one that is swift, swift, slow. Galen emphasizes his own contribution and ability to judge this subject by virtue of long, hard work at learning to feel the phenomena. He did not learn it from his teachers because they did not know it (K 8.786–788). His model, Archigenes, used terminology too vague to describe the phenomena, Galen says. I have found no scholar who has tried to judge Galen's claims.[32]

Galen's large work *Therapeutic Method* (K 10.1–1021), usually called *Methodus Medendi* in Latin, often *Megatechne,* makes the most reasoned claims to follow the science of Hippocrates: Hippocrates did not, indeed, work out the science in detail. He omitted much, especially regarding complicated conditions. But he did show the way, and his system is the only basis for a method of healing. None of Galen's predecessors completed the work: some were totally ignorant, some who knew it were unable to add what was needed, some chose to conceal and obscure it (K 10.632–634). Galen himself will, he says, write the first good

---

31. The major works and their location in Kühn's edition: *Use of the Pulse,* 5.149–180; *Distinction of Pulses,* 8.493–765; *Diagnosis by Pulses,* 8.766–916; *Causes of Pulses,* 9.1–204; *Prognosis from Pulses,* 9.205–430.

32. Galen's description of his own research has been paraphrased at length and discussed by Karl Deichgräber, *Galen als Erforscher des menschlichen Pulses,* Sitzungsberichte der deutschen Akademie der Wissenschaften zu Berlin, Klasse Sprach, Literatur, und Kunst (1956), no. 3. For a helpful exposition of Galen's ideas about the pulse, see C. R. S. Harris, *The Heart and Vascular System in Ancient Greek Medicine* (Oxford, 1973), pp. 397–431.

work on therapeutic method ever written. The title, *Therapeutic Method* or *Method of Healing* or *Method in Medicine,* is polemic in tendency. Galen is taking away from the Methodists their claim to have method, saying that they have no knowledge of how many diseases there are and no analytical techniques. The Empirics are virtually as badly off. The Empirics are not the opponents in this work, except that when the Methodists are refuted, the Empirics might move in to claim the territory. Hence Galen's primary attack is on the Method as presented by Thessalus, which is full of false and meaningless claims; secondarily, he attacks the Empirics because they have no way of finding correct treatments. To both he opposes his own, which he calls Hippocrates' method: there can be no precise cures in books, no specific medicines for diseases, only the nature of the patient and his disease which the physician readjusts by contraries (K 10.172–183).

At the beginning of book 7 Galen explains that he had broken off the project when the addressee of books 1–6, Hieron, had died, but now, fifteen years later (probably after 193 A.D.), he will complete his design, not to gain fame, but as a favor to his friends (K 10.456–458). The design of the whole, completed in fourteen books, gives an account of diseases, divided into three categories: those that involve separation of tissues (wounds, ulcers, ruptured vessels, fractures); those from *dyscrasia,* that is, humoral imbalance, in the whole body; and those that involve unnatural swellings (inflammation, tumors, cancers, and the like). In all diseases one must distinguish cause from affection and symptoms, treat the cause, and restore the body to its natural condition. All internal and external lesions are comparable to a pinprick: if the body is healthy and balanced, *eucratos,* the tissue will unite (K 10.386–401). If there is *dyscrasia* in the body, the wound will fester because of the body's excrement, which will keep the lesion from closing, and one must treat the *dyscrasia* as well by purge, diet, bloodletting, and medicine that restores the lesion to its proper heat or coolness, moistness or dryness. All lesions everywhere require the same approach, but every individual has his own natural temperament (*krasis*), which you should know from observation, and every kind of tissue and area of the body has its own temperament: the ear lobe differs from the hand or trachea; a nerve is much drier than a muscle and needs more drying factors in the medicines, many of which Galen claims to have invented by using the proper logic (K 10.394–395).

Hence his method requires insight into the precise mixture of hot, cold, wet, and dry in every foodstuff and drug, every part of the body, and every individual, along with how and when to treat with contrary elements to restore the natural condition. There are no "communities" of disease as the Methodists claim, nor can discoveries be made empirically: one must know by reason precisely the effect needed and the way to achieve it, besides allowing for seasons, times of day, and any other complicating factors (cf. K 10.31–38, 162–173, 316–357). Galen claims that he has always succeeded where Methodists had failed, taking over their cases and exposing their murderous ignorance—they think a medicine for an ulcer would be good for any person and any part (cf. K 10.204–205, 390–391). With fevers (caused primarily by putrifying humors), one must know where humors are lodged and what their consistency is, then must thin the thick ones, warm the cold ones, flush them out, bleed them out, and must do it before the disease gets worse, making the proper adjustments for all variables. For unnatural swellings, one must treat locally and treat the *dyscrasia* of the whole body by purge and bleeding and by diet that warms or dries. These principles, treated with great detail of phenomena he has observed, experience from his own practice, and polemics about others' stupidity, make up Galen's *Therapeutic Method*.

Galen's literary model, insofar as he has one, appears to be the *Therapeutic Method* of Thessalus, the manifesto of the Methodist school, which taught how to recognize the "communities" or common conditions in diseases and to restore the body to its original symmetry or harmonious relationships in the sizes of the pores, concepts borrowed from Themison and Asclepiades and developed by Thessalus with new terminology (K 10.7–8, 250–275). Thessalus is particularly odious to Galen because he criticized all physicians who preceded him and said that Hippocrates had left a harmful heritage—he even criticized the

Aphorisms. Galen responds by quoting Plato, from the *Phaedrus,* on Hippocrates' *methodos* (K 10.13) and in the work periodically returns to the metaphor of the road (*hodos*): Hippocrates pointed out the straight road, which all good philosophers and physicians have followed (K 10.117), but nobody completed it. Galen is like Trajan, who found the roads of Italy often muddy, rough, and difficult, but left them all smooth and passable (K 10.632–634).

Galen had preceded the writing of *Therapeutic Method* with a briefer version of the subject in two books, *Therapeutics to Glaucon* (K 11.1–146). He followed it much later with *Affected Places* (K 8.1–452), which treats the subject of finding the source of the disease by study of the functions that are impaired. His literary model for *Affected Places* is the Pneumatic Archigenes' work of the same title, which Galen corrects in details: Archigenes indulged in useless speculation on how causes should be named and claimed that species of pain could be elaborately classified and named as indications of parts affected (K 8.90–120). Galen says that attention to Hippocrates would have saved Archigenes from aberration. For example, Archigenes could not have said that "heavy pains" indicate the liver had he noticed and thought properly about Hippocrates' statement (*Epid.* 6, L 5.268) that kidney pain is heavy (K 8.113). But Galen has better, more logical arguments as well: no one can have felt all the pains and know them; language for categorizing pain is imprecise and in the end not useful (K 8.113–120).

The center of Galen's developed medical system, the anatomy and physiology, are laid out in three major works, *Anatomical Procedures, Use of the Parts,* and *Natural Faculties.*[33] In each Galen makes an almost ritual gesture toward proving the science Hippocratic. *Anatomical Procedures,* Galen tells us, is original in very

33. *Anatomical Procedures,* partly in Greek in Kühn, vol. 2, partly in Arabic, ed. Max Simon (Leipzig, 1906). The whole is available in English translation by Charles Singer, books 1–9 (London, 1956), and W. L. H. Duckworth, the later books, 9–15 (Cambridge, 1962). *Natural Faculties,* Kühn, vol. 2, *Scripta Minora,* vol. 3, is available in English in the Loeb Classical Library translation by Arthur Brock (London, 1916).

large part (see *Scr. Min.* 2.102–108 and *Anat. Proc.* passim) and superior to manuals by his predecessors, the Alexandrians and Marinus. Galen says that Hippocrates must have known all about anatomy, but refrained from writing about it because anatomical instruction was oral and kept within the family (*Anat. Proc.* 2.1, K 2.280–283). Galen reordered Marinus' presentation while adding his own numerous observations and, unlike Marinus, separated the anatomy from the physiology. In the physiological work, *Use of the Parts,* which parallels the Anatomy part by part, beginning with the hand, Galen used Aristotle's *Parts of Animals* loosely for his model. (I noted above that he conceived this eloquently teleological work in connection with the Peripatetic circle in Rome and dedicated book 1 to Boethus; also that beyond the first book he does not attempt to prove the science to be known by Hippocrates.) In the *Use of the Parts* Galen's opponents, whom he wishes to put down, are Erasistratus and Asclepiades. Asclepiades appears to have eschewed teleological explanations entirely and so becomes a natural opponent. Erasistratus seems to have attempted explanation of nutrition and excretion by notions of pressure, vacuum, and congruence of material with passages. At times he confessed his ignorance (for example, about the function of the spleen, an admission that infuriated Galen) and felt that Nature could have done better than she did (cf. esp. *Nat. Fac.* 2.4, 2.6; *Scr. Min.* 3, 165–171). Galen's eloquence on Nature's providence needs Erasistratus' doubts for a target. Erasistratus' work *General Principles (katholôn logoi)* is perhaps the literary model as well as the opponent for Galen's *Natural Faculties.* Erasistratus had hoped to explain natural actions in the organism and how they occur (cf. K 2.63–65). Galen replaces those explanations with notions that the organs have faculties to *attract* what they need, *assimilate* it to themselves, *hold* it meanwhile, and *expel* the unneeded remainder or excrement. Galen gestures toward proving this science Hippocratic by twice quoting from *Airs Waters Places* 21: "The mouth (*stomachos*) of the uterus cannot attract (*eirusai*) semen" in a pathological condition (K 2.61, 2.187). Either he is misquoting from memory, as I suspect, or there are textual problems: the

Hippocratic manuscript tradition offers *stoma* (mouth) and *hypodechetai* (receive); the mouths of Scythian women's uteri, clogged with fat, cannot receive seed.[34]

Galen's rhetorical posture in relation to Hippocrates is thus partly a superficial denial of what he was obviously doing as he covered virtually all areas of medicine and imposed his personality on them: he took from all sources what he considered best and made it his own by improving on logic and classification and by fitting it into his own comprehensive system. Because he considered his direct sources wrong and incomplete, he needed to give them only as much credit as he wished, for he was Hippocrates' direct heir. His argumentative tactic of selecting an enemy and maneuvering him into the position of being an enemy of Hippocrates reduced his problems of sorting out his real intellectual and technical antecedents, but has caused problems for the history of medicine because Galen survived to give testimony and the others did not. At different times Galen has seduced people into crediting him whole, as Hippocrates' direct heir, or half, taking Hippocrates as the source of true science perversely misunderstood by those after him. Galen's rhetorical posture would seem to gain credence because in a sense it reflects his real situation: to a great extent he was a unique genius and the unique prophet of his version of medicine and of intellectual history.

The form and substance of a single major work, the *Hygiene*, demonstrate the method in Galen's major works: his use of literary sources while triumphing over them and his concoction of a historical amalgam, which he called classical, in response to contemporary concerns. The *Hygiene* and the accompanying work, *Thrasybulus, Whether Hygiene is a Part of Medicine or Gymnastic*,[35] were written toward the end of his series of major works, at about the time he began his Hippocratic commentaries.

---

34. Cf. Hans Diller, *Hippocratis De aere aquis locis*, CMG, 1.1.2 (Berlin, 1970), p. 72. For another apparent misquotation to prove the same point, cf. K 2.189, K 2.196.
35. The *Hygiene* (*De Sanitate Tuenda* K 6.1–452) was edited for the Corpus Medicorum Graecorum by Conrad Koch in 1923(CMG 5.4.2) and is cited by page from his edition, as well as by book and chapter. It is available in

Thrasybulus takes up the current question Galen's friend Thrasybulus has asked him to treat: does hygiene belong to medicine or to gymnastic? In his usual fashion, Galen deprecates quarrels over words, while showing his own adeptness in such quarrels. After demonstrating, in quasi-Platonic manner, that there are many ways one can logically divide the subject of care of the body and many terms one can use, Galen concludes that there is a single comprehensive science of care of the body, which is divided into therapy for the ill and maintenance of health in the healthy (pp. 90–92). Having made his logical point, Galen proceeds to a diatribe against the ignorant class of men, who have been trained as swinish athletes and are useless to society. But Plato and Hippocrates both say correctly that gymnastics as training for athletes has nothing to do with proper hygiene. He quotes Plato from the *Republic* (pp. 80–82) and, to prove that Hippocrates held the same view (and that Plato followed him), he quotes *Nutriment* (34; L 9.110): "The athlete's condition (*diathesis*) is not natural, the healthy state (*hexis*) is better" (p. 83). In his usual manner, Galen calls a roll of noble figures, "followers of Hippocrates," who thought as he did: Diocles, Praxagoras, Phylotimus, Erasistratus, and Herophilus, "students of the whole science of the body" (pp. 85, 99), who are to be contrasted, for example, with Theon and Tryphon, whose topics show that they are concerned with the vile science of athletics: they talk of "preparation for exercise," "partial" and "total" exercise, "recovery" (*apotherapeia*), and so on (pp. 99–100).

Galen gives some hints as to why Thrasybulus requested that he write. In fact he brings the controversy to life on the streets of Rome:

> The healthy city hates and despises this activity which perverts all one's force for living, and turns it into an unworthy condition of the body. I have often found myself stronger than highly respected athletes. They were useless when it came to traveling or

English translation by Robert Montraville Green, *Galen's Hygiene* (Springfield, Ill., 1951). *Thrasybulus* (K 5.806–898) was edited by George Helmreich for *Scripta Minora*, vol. 3, and is cited by page from that edition.

military activity, and more so in political life and farming. And if
it is necessary to stay with a sick friend, they are totally worthless
for consolation, consultation, or assistance—just like pigs. Nevertheless the most unsuccessful of them, who have never won a
prize, of a sudden set themselves up as trainers, and cry their
wares loudly as swine in their barbarian voices. Some even try
writing, about massage or conditioning, or hygiene, or exercise,
and dare to take on and contradict people they don't understand
at all. Like that man the other day who criticized Hippocrates and
said he did not know about massage. When I came up, some of
the physicians and philosophers who were there asked me to explain the theory to them. I made it clear that Hippocrates was first
in the subject and had said all that was best. But that self-taught
trainer suddenly entered the fray, stripped down a slave boy and
told me to massage and exercise him or to be silent about massage
and exercise. And then he bawled, "When did Hippocrates enter
the wrestling ring, or the wrestling school? He didn't even know
the art of the oil rub!" So he kept shouting, and would not be
quiet and listen to what was said. But I calmly explained to those
who were there that the crude fellow was like a baker or cook who
boldly discourses about barley broth or bread. Then I said,
"When did Hippocrates work in a cookshop or at a mill? Let him
be a maker of pastries, bread, sauce and fish, since only so should
he talk about them." [*Thrasybulus* pp. 96–98]

Galen seems quite pleased with his witty retort to the trainer.
His vengeance on the barbarous fellow and his tirade is not
confined to the *Thrasybulus,* but extends to the *Hygiene.* In the
Hygiene, which seems to have been written shortly after the
Thrasybulus, Galen fulfills the program he had laid out for the
science of hygiene, treating health in very broad terms and ad-
dressing himself to all conditions of life. The result, as in much
of his work, is somewhat marred by haste, repetitiousness, and
failure to fulfill the design entirely but it has justly been much
admired. One can see in the structure of the writing how Galen
took over the work of Theon and corrected and improved it by
imposing on it his own design and his own science.

In book 1, Galen lays down the principles of the science of
hygiene as he sees them. Hygiene is the part of medicine that
deals with preserving health. To practice it properly, one must

know what health is. One must know that from the basic elements, fire, water, earth, and air (or, equally, hot, cold, wet, dry, it makes no difference, 1.2, CMG 5.4.2, p. 4)[36] are produced *homoiomerê*. Health can be described as organic functioning according to nature. Health is a mean between extremes, relative always to the natural condition of the particular organism. The science of hygiene provides regimens that will prevent or cure *dyscrasia*—excess of hot, wet, cold, or dry—in particular parts and in the body as a whole.

In six chapters of book 1, Galen thus gives a succinct summary of his physiological principles, with numerous cross-references to his previous works, which the hygienist should read. He then lays out the structure of his presentation: he will describe the care of the healthy body from birth to old age and then discuss hygiene in diseased conditions. He deals summarily with the child from birth to seven years, giving advice about breast feeding, bathing, massage, exercise, and pure water and air. For the second seven years, he says simply to continue the regimen of the first, without violent exercise and with warm baths. At this point (chs. 12-14), he explains his conception of the excrements (*perittômata*) of bodily processes.

In books 2, 3, and 4, Galen treats the kinds of exercises and their effects, preparation for exercise, treatment after exercise (*apotherapeia*), massage, diet, bathing, and other details relating to exercise and fatigued conditions. He seems to be following Theon's work and mentions him frequently. He even says that Theon is the best who has written on the subject (p. 53).[37] Book 5 deals in summary fashion with old men's problems: the best exercises, foods, and wines, remedies to use in respiratory infections, how to keep the intestines functioning. Galen also offers

36. Although Galen frequently mentions the triumph of his philosophical distinction—hot, wet, cold, and dry are not basic elements, but fire, water, earth, and air—and refers readers back to his *Elements according to Hippocrates* for proof that Aristotle (or Plato) followed Hippocrates, he as frequently says the distinction makes no difference to medicine or neglects to mention it at all.

37. Whether other sections of the *Hygiene*, such as hygiene of nursing, gerontology, and wines, are closely modeled on predecessors is not apparent. I have found no direct sources.

examples of special attentions required by peculiar bodily types. Book 6 gives assorted advice and recipes for serious problems of indigestion and ill health, with many suggestions that the reader pursue the subject in Galen's other works.[38]

Galen's achievement, of which he boasts repeatedly, is that he has written a scientific hygiene for everyone, not only for the perfect body in its prime. Sketchy as the work is in many places, it successfully completes the design conceived in the first book, and it presents a coherent theory throughout. Galen attains his usual rhetorical success as well, not only using Theon, but showing his inferiority because he did not understand Hippocratic science. His manner is instructive for purposes of assessing the Hippocratic tradition and Galen's peculiar construction of it.

Galen makes some desultory gestures toward proving that hygiene is a Hippocratic science. He quotes "Opposites cure opposites" (pp. 17, 159, 163, 167); "Purge what is concocted, not raw" (p. 116); "Spontaneous fatigue is a sign of illness" (p. 104); "Exercise should precede food" (pp. 39, 141, 163); and some few other aphoristic hints from which he concludes that Hippocrates understood the science as he himself constitutes it.[39] Galen's most successful quotation, however, with which he thoroughly humiliates Theon, is from *The Surgery* (L 3.322): "Rubbing can relax, tighten, increase flesh, attenuate. Hard rubbing tightens, soft relaxes. Much attenuates, moderate thickens."[40] When, in book 2, Galen reaches the subject of massage, he talks of trainers who pretend to know more than Hippocrates, although no one has ever said more about massage than Hippocrates did (re-

38. I see no reason to treat book 6 as having been written at a considerably later date, as Bardong does (p. 639). Its citation of works written ca. 182 seems most likely to have been added when the last chapter of book 6 was tacked on.

39. The index in CMG 5.4.2 gives the list. Treatises on which Galen draws are *Aphorisms, In the Surgery, Nutriment, Humors, Use of Liquids, Epidemics* 6, and *Breaths* (only for "Opposites cure opposites," which Galen attributes to *Aphorisms*, p. 163). "Exercise should precede food" is also falsely attributed to *Aphorisms* (pp. 39, 143, 163).

40. I have been tempted to think that Galen found this quotation after the incident he describes in *Thrasybulus* since he does not quote it in that work. But there is no way to be certain, and it may seem unlikely on the face of it that the passage was not long part of Galen's mnemonic system.

minding us of Galen's occasion for irritation in *Thrasybulus*). Theon, Galen writes, says,

> Those who write about massage believe that one should always harmonize quantity.... Soft massage is productive of three results according to quantity: a little relaxes the flesh and makes it soft to the touch, a lot wears it out and melts it, while enough fleshes out the body with fluid, flaccid flesh. Similarly, hard rubbing has an equal number of results depending on quantity. A lot binds up the body, constricts it and causes something equivalent to inflammation, enough fleshes up the body with long-lasting, well-formed flesh, while a little makes the superficies red for a short time." [Ch. 6, p. 44]

Galen retorts that Theon is saying nothing, or, at best, repeating what Hippocrates said. Galen finds Theon vague or unintelligible (Theon said *anienai;* Hippocrates' term was *lusai,* pp. 44–45) and clearly inferior to Hippocrates. There is more to come, however. Galen next infers that Theon clearly either could not understand Hippocrates or did not wish to praise his brief statement (ch. 7, p. 47). One might think that Hippocrates mentioned only four items, but no, the extremes imply the means, and therefore Hippocrates talks of nine qualities and quantities (three times three, for which Galen offers a diagram), whereas Theon is so foolish as to mention only six explicitly. Finally, in high dudgeon, Galen accuses Theon of slandering Hippocrates (p. 50) by depriving him of his just praise and perverting his views. Gentle Theon, who, as Galen quotes him, did not claim originality, has become another enemy of Hippocrates. After giving his own, more scientific breakdown of massage, Galen explains what the Hippocratic method is, thus completing his street quarrel with the vulgar trainer. Here is Galen's version:

> This is characteristic of the method: the ability to proceed from brief elementary principles to all the separate details, and to judge all erroneous statements by comparing them to scientific standards, as though to a canon. This has been enough to show that no one had a correct knowledge of massage, not even the gymnast

Theon, despite his being superior to others, except for Hippocrates and those who follow Hippocrates. [Ch. 4, *fin*, CMG 5.4.2, p. 53]

Later in the *Hygiene*, Galen again puts Theon down severely. In book 3, where the subject is treatment of fatigue, Galen notes that Theon implies that bathing is good for fatigue. Galen produces another diagram: there are four simple kinds of fatigue, which with their combinations make fifteen kinds all told. And not all require bathing (ch. 8, pp. 92–95). Besides, Galen again pretends to find Theon's language unintelligible, and, to show the source of Theon's problems, Galen again quotes him. The man proceeded from experience, not theory: as he admits: "If this has a theory (*logos*) underlying it, that is fortunate. If not, the evidence of the results is not to be accepted unless it has a concurrent theory in good time" (p. 93). Again, Theon sounds reasonable, even when quoted out of context, but we cannot judge. Galen's purpose as scientist is to replace Theon's uncertainty with an exhaustive, logical scheme.

One more tendency in Galen's writing exemplified by the *Hygiene* is the Greek humanism, which could be called chauvinism depending on its manifestation. The science of guarding health is naturally a vehicle for recommending moderation in all things: eating, drinking, exercise, work, sleep, and so on. Galen presents it also as peculiarly Greek. When he is discussing the care and bathing of infants, Galen glances aside at the barbarians:

> The Germans do not raise children properly. But we are not writing for Germans or other savage or barbarous people, no more than for bears or lions or other wild animals, but for Greeks, or those born barbarian who imitate Greek ways. Who among us would carry his warm newborn babe to the river and there, as they say the Germans do, test his nature and strengthen his body by dipping him in the cold water like heated iron? ... An ass or any mindless beast might get the greatest benefit from having a skin thick and hard enough to bear the cold without suffering. But for a human, a reasonable creature, of what significance could it be? [1.10, pp. 24–25]

I have quoted only a small part of what Galen says on the subject. Here, too, he uses an obscure quotation from Hippocrates to prove his point: "Porosity of body for transpiration: healthier for those who lose more, unhealthier for those who lose less" (*Nutriment* 28; L 9.108). This may mean "Nothing too much" as regards porosity, which is Galen's interpretation (p. 25), or "the more the better." When describing the effect of climate and geography on physique, Galen says that the perfect physical type and the perfectly healthy body are possible only in the proper climate, which to be perfect must approximate that in the homeland of Hippocrates. One would not find it among Northern Europeans, Africans, or Asians (2.7, p. 56).[41] The perfect man is the one imaged in Polycleitus' perfectly proportioned statue called the canon (p. 56). Again, when Galen outlines his Hippocratic method, he does so in the image of the canon: the Hippocratic method reduces everything to absolute essentials, to the brief elementary principle from which one can judge all erroneous statements "by comparing them to scientific standards as though to a canon" (2.4, p. 53). When Galen refuses to understand Theon's language but finds Hippocrates clear and precise, he is obviously behaving like the Parisian who finds outlandish French unintelligible.[42] Against barbarism of all kinds, including that of the vulgar trainer, Galen allies himself with the classical age of Hippocrates and Plato, and he can always call a roll of

41. Galen alludes, as he does frequently, to the Hippocratic treatise *Airs Waters Places*, though he never mentions it in the *Hygiene*.
42. Galen wavers in his rhetorical postures in this matter. He frequently presents himself as one who cares nothing for quibbling about language as long as the matter is clear, and he says that Plato agrees with him (K 2.581, cf. De Lacy, "Galen's Platonism," pp. 29–30). He wrote against the quibbling Atticists of his time (*Scr. Min.* 2.90). Scholars have tended to generalize from some few of Galen's statements without considering the posturing, indeed the chauvinism, involved. Galen insists that, although he himself is cultured, he can be generous. It is not people's bad grammar but their lack of precision that he detests: let them say *"phere ton pous"* if they wish (K 8.589). But he finds them imprecise as well. We do not know how Galen put down the Atticists, but I suspect that he found them imprecise, inaccurate, and quibbling, and he likely did so in the name of Hippocrates, some of whose works (as Galen used them) are not up to classical standard. *Nutriment* is a good example. Galen sometimes had to put on blinders when he indulged his faith.

physicians and philosophers in whose tradition he stands against any current opponent. But each of the physicians and philosophers, save Hippocrates and Plato, may be the opponent elsewhere and be subjected to a similar roll call of their betters.

Throughout this period, despite all of his claims for Hippocrates as his guide and ideal, Galen shows very little knowledge of or concern about the transmission of the Hippocratic writings or of others' views about them. He is a student of Hippocratic science in his own version; but who had shared that particular version or what other versions there were is not revealed by him, nor does he acknowledge that there are problems related to the Hippocratic texts, except in the few instances we shall now consider. They are evidence that in his early career Galen was very vague on the subject of the tradition of the Hippocratic Corpus, but began around 175–177, his forty-fifth through forty-seventh years, to become more knowledgeable.

I have mentioned Galen's assertions in *Opinions of Hippocrates and Plato* of the spuriousness of the later parts of *Nature of Man,* along with his inconsistent citation there of a sentence from the "spurious" portion as a sentiment of Hippocrates, as well as his general acknowledgment that there are Empiric traditions of Hippocratic interpretation (cf. *Affected Places* 3.3, K 8.141–142. The Empirics simply treat pneumonia as their teachers did, or as Hippocrates does in *Regimen in Acute Diseases,* and do not inquire where the infection is.). One passage in *Therapeutic Method* suggests that a Methodist defending the Method of Thessalus cited the Hippocratic treatise *Wounds* (L 6.402) to illustrate the process of making an old, unhealed wound new to facilitate healing. Galen argues at length against that (fairly sensible) interpretation of the text, citing many other passages from the treatise to prove that he understands it (K 10.277–286). But it is very doubtful that Methodists often adduced Hippocrates in their support. The little work *Regimen in Acute Diseases according to Hippocrates*[43] is an answer to an unspecified Methodist physician who wrote that Hippocrates fed fever patients daily. Galen

43. *Galeni De diaeta Hippocratis in morbis acutis,* ed. and trans. Malcolm Lyons, CMG Suppl. Or. 2 (Berlin, 1969).

wrote for his noble friend Victor, who had asked him about the Methodist's statement. The Methodist is wrong, of course, as Galen proves by paraphrase of *Regimen in Acute Diseases*. Erasistratus was wrong, too, when he slandered Hippocrates' students Dexippus and Apollonius by saying that they gave only three tiny ladles of water a day to fever patients. Galen refutes another statement by the progressive Methodist: *chondros* was not available in the time of Hippocrates, else he would have used it. Galen retorts that it certainly was available and cites *Regimen* as proof, adding: "Your friend, of course, claims that the work is not by Hippocrates, and if you like you can take its author as Philistion or Ariston or Phaon. All of them without exception are ancient" (p. 109). Poets also mention it, says Galen, and so does *Affections*, of which "the author is either Polybus, the most famous of Hippocrates' pupils, or Euryphon, who was a famous man and contemporary of Hippocrates" (p. 109). Galen's concern here is chronology: he does not want the Methodist to say that Hippocrates was ignorant of a good therapy that developed later. But his specifications about authorship of Hippocratic treatises belong to a new interest in the subject that he developed around 175 A.D.

On Coma according to Hippocrates (CMG 5.9.2, 181–194)[44] is a short treatise that asks how Hippocrates used the word and answers that he used it for a lethargic state that could be accompanied by restlessness. Galen does not give his purpose for writing the treatise, but the direction of his argument indicates that he was disputing the interpretation of the meaning of the word in *Epidemics* 3 that had been offered by "Hippocrateans" (unspecified). They had interpreted "coma and then wakefulness" as a mysterious condition they call *typhomania* (pp. 183–188). To prove his point about the meaning of *coma,* Galen quotes the opening of *Prorrhetic* 1, where it is clearly stated that the coma and sleeplessness go together (p. 181). *Prorrhetic* 1 asks, "Are they phrenitic?" To illustrate that Hippocratic stylistic mannerism, Galen quotes from *Epidemics* 3 and 6 and from another

44. *De comate secundum Hippocratem,* ed. Ioannes Mewaldt, CMG 5.9.2 (Leipzig, 1915).

part of *Prorrhetic* 1 (p. 192). Galen clearly considered *Prorrhetic* to be a genuine work of Hippocrates when he wrote this treatise. Yet he acknowledges a problem: after quoting *Prorrhetic* 1 for the first time, he quotes *Epidemics* 3 "to get testimony from works agreed to be genuine" (p. 181). Galen does not mention, if he knows it, that the term *typhomania* actually comes from *Epidemics* 4 (L 5.105). Galen later changed his mind about *Prorrhetic* 1, when he was irritated at Lycus' empiric use of it, but he never reveals the occasion for the dispute on *coma,* nor the identity of those who might not agree that *Prorrhetic* was not genuine.

The work *Faculties of Foods,* which fulfills part of his general system by treating the *kraseis* of foods that have to be related to the temperaments of patients, was written after the *Hygiene* (cf. CMG 5.4.2, p. 252), near the time of his earliest Hippocratic commentaries, ca. 175–177 A.D. Galen draws on predecessors' descriptions of foodstuffs and their effects (primarily works by Diocles, Mnesitheus, Phylotimus, and the Hippocratic treatise *Regimen*) and for each food gives some account of its *krasis* that fits the theory. I have quoted above his account of the Hippocratic *Regimen* in this work (p. 59). It seems clear that his various accounts of the same work in the period 175–177 show an increasing attention to authorship and an increasing precision. The accounts, in apparent sequence, are as follows:

1. The treatise on *Regimen in Acute Diseases according to Hippocrates,* mentioned above (p. 115).

2. In the commentary on *Regimen in Acute Diseases* (CMG 5.9.1, p. 135), where he calls it *Regimen in Health,* Galen says "they" attribute it to many authors—Euryphon, Phaon, Philistion, and Ariston—but it belongs to the great Hippocrates, son of Heraclides, whose grandfather, Hippocrates, son of Gnosidicus, wrote nothing according to some and only *Fractures* and *Joints* according to others.

3. In the commentary on the *Aphorisms* (K 18A, p. 8), where he calls it the *Healthy Dietetic,* Galen says that it is attributed to Hippocrates, but "they deny it to him": some attribute it to Philistion, some to Ariston, some to Pherecydes.

4. In *Faculties of Foods* (CMG 5.4.2, pp. 212–213), Galen says that it is ascribed to Hippocrates by some and also to Philistion,

Ariston, Euryphon, and Philetas. The second book is worthy of Hippocrates, but the first book (the philosophical schema) departs from his views. Transmitted as a whole it is called *On the Nature of Man and Regimen*. The second and third books transmitted alone are called *Regimen*.

5. In the commentary on *Epidemics* 1 (CMG 5.10.1, p. 108), he refers the reader to his writing on dreams: "I have written on dreams elsewhere, including those which indicate bodily disposition as described in the book *Healthy Regimen*" (that is, the extant *Regimen*, book 4). Galen's book on dreams has been lost, save perhaps for a small fragment (K 6.832–835).

These various statements indicate that in the process of writing *Faculties of Foods*, where he uses the treatise extensively, Galen corrected his earlier rather careless claims by reading book 1 and deciding that he had to agree with "them" that the work could not be by Hippocrates because it does not use four humors. That it judges foods by hot, cold, wet, and dry and *kraseis* nevertheless makes it peculiarly congenial to his method. Who "they" are who give the putative list of authors cannot be known for certain, but I shall suggest below (p. 239) that the list goes back to Dioscurides and Capiton.

In his work on *Trembling, Shaking, Spasm, and Shivering*,[45] Galen quotes *Sevens* for Hippocrates' explanation of different classes of diseases that arise from *dyscrasia* in the body's innate heat (too much heat, too much cold, [K7.618]). The Hippocratic rationale for *rhigos* which Galen gives is based on that passage. In his work on *Marasmus* (withering away),[46] however, when he

---

45. Latin title *De tremore, palpitatione, convulsione et rigore* (K 7.584–642). The work completes the scheme of the Causes of the Pulse, assigning causes to all involuntary bodily movements. Praxagoras, who found the cause of all in the arteries, is the literary opponent (K 7.584). Galen's original contribution in the work seems to be insistence on distinguishing the terms *tromos* (trembling) and *rhigos* (shivering), which Athenaeus had not done (K 7.609–610). Athenaeus' work is probably his literary model.

46. Latin title *De marcore* (K 7.666–704), written at the same time as the *Hygiene* (Ilberg 3, 188, as corrected by Theoharis [who translated the treatise in *Journal of the History of Medicine and Allied Sciences* 4 (1971), 369]), therefore near the work mentioned in the previous note. Its subject is the cause and cure of withering in pathological conditions and in old age.

is discussing and refuting the view of "almost all recent philosophers and physicians" (K 7.674) that the body's heat increases until its prime and is the cause of withering, Galen cites the passage from *Sevens* which someone had apparently used to support or illustrate that view: "The heat that produces our bodies also kills us" (K 7.675, cf. L 8.644). Galen's response to the quotation is, "First we shall say, my fine fellows, that it is not one of the genuine books of Hippocrates in which that is said. And second, if it is a belief of Hippocrates they should explain what he means and offer some demonstration" (K 7.675–676). The question of the tradition of Hippocratic interpretation that Galen rejects can perhaps be illuminated by a passage in his later commentary on *Epidemics* 6 (CMG 5.10.2.2, pp. 270–274). The passage in question reads: "Man's soul grows continuously until death. If he is fevered both soul and body are consumed by the illness" (L 5.314). Galen denies that the statement is intelligible or genuine because one would have to know the substance of the soul to understand it, and no such doctrine can be found in Hippocrates' genuine works. But, in giving various philosophical views on the subject, Galen speaks of the Stoics' view that the soul is a dry pneuma that needs air and nutriment. "And those who think Hippocrates was guide (*hegemon*) in that doctrine, as it is given in *Sevens*, say that 'grow' here means addition of those two substances" (CMG 5.10.2.2, p. 273). In the work on *Marasmus*, Galen generally follows Archigenes and Philip, pneumatic writers (cf. K 7.685–689), while rejecting various of their results. In *Trembling, Shaking, Spasm, and Shivering*, he follows Athenaeus, the Pneumatic, at least in part (cf. K 7.610–615). In this last work Galen found it convenient to accept the view Athenaeus attributed to Hippocrates without questioning the genuineness of the work (*Sevens*) from which it came. In *Marasmus*, Galen could not accept the doctrine and rejected *Sevens*. Galen's source for the Stoic interpretation of the passage in *Epidemics* 6 is not clear, but it may be Aephicianus, his teacher.[47]

---

47. Galen's sources for the views he quotes in his commentaries and Pneumatics' quotations from Hippocratic texts will be discussed later. Here I might note that Athenaeus, when talking about trembling and shivering, quoted a sen-

In short, Galen varies between carelessness and pedantry in attributing material to Hippocrates, depending on the needs of his argument. He is always concerned that "Hippocrates' " view accord with his own, but is not concerned with consistent attributions of Hippocratic works. Thus, before ca. 175 A.D., Galen indicates that he knows there is scholarship on Hippocratic texts, but does not have much knowledge of it or pay attention to it.

The work on *Abnormal Breathing*[48] is pivotal in Galen's literary career and his Hippocratism, for in its writing he claims authority not only in Hippocratic science, but in the texts as well. Claims made in this work are followed up in the Hippocratic commentaries that were begun soon afterward. The work itself is an attempt to catalogue, for diagnostic purposes, variations of breathing from the normal, to show that Hippocrates knew them all, and that all that Hippocrates says on the subject is right. Galen's procedure is similar to that in his work on pulses: indeed he works out the analogy with the pulse as far as he can, since he conceives the operation of pulse and breathing in the same way. The lungs, like the arteries, expand to inhale, pause, contract for expulsion, pause, and so on. Depth, rate, and rhythm of breathing as they vary from the normal indicate pathological states. Increase of need for cooling, for expulsion of products of combustion from the heart, or for renewal of psychic pneuma causes increased depth and rate; decrease of such needs does the opposite (pp. 765–773). Pain in the thorax will decrease the depth (here unlike the pulse, pp. 774–778). Peculiar states produce peculiar patterns: in sleep more is exhaled than inhaled to get rid of smoky excrement from digestion (p. 772); muscular problems or weakness of controlling power cause anomalies, as does mental aberration (pp. 803–

tence from Plato on the subject, without, apparently, giving the source of the quote or naming Plato. Galen in response quotes the whole passage from the *Timaeus* to show that he knows where it comes from (K 7.610). "Hippocrates" may well have been quoted in the same manner by Pneumatics, but I can find few likely instances.

48. *Peri dyspnoias,* Latin title usually *De Difficultate Respirationis* (K 7.753–960). Kurt Bardong was preparing an edition of it for the Corpus Medicorum Graecorum in the late 1930s. His article dates the work to the beginning of 175 A.D.

809). Deep but infrequent breathing is a sign of mental aberration (pp. 789–791, 809; the word he uses is *paraphronesis*, his examples are delirium, pp. 826–832).

Galen claims originality in this subject: no one has worked out the science of diagnosis by breathing, some ignore it entirely, some barely mention it. Hippocrates' *Epidemics* should have waked them up, but did not, even though people call themselves Hippocrateans and write exegesis of his works (pp. 763–764). There also seems to be a social stimulus for this work, as for others. Someone (another of Galen's shamelessly jealous opponents) pointed out a patient who was delirious but whose breathing was not deep and infrequent, "and he laughed, nay he guffawed shamelessly, ridiculed Hippocrates and me, and went away" before Galen could get in a word and give him proper instruction (pp. 833–834), namely, that the abnormality is a sign of mental aberration, but does not always accompany mental aberration. (Galen points out in *Epidemics* 3 cases where other kinds of breathing are mentioned, pp. 835–847. All his examples of mentions of the specific aberration come from *Epidemics* 1.)

After his exposition of the science of abnormal breathing in book 1, Galen devotes two books to his demonstration that Hippocrates knew the science and was right. His demonstration requires a considerable amount of special pleading, which produces a refined description of Hippocratic science: why in many cases does Hippocrates not mention abnormal breathing where it must have occurred? The answer is that in his brevity he does not mention what would be obvious to all specialists. His is technical work for sophisticates (pp. 850–854). If one is properly instructed in the science one can see what he means. Thus, when deep, infrequent breathing with delirium is not mentioned, either it should be obvious to all (p. 876, Erasinus in *Epidemics* 3), or wasn't noticed because the patient was so wildly delirious (p. 877, Criton), or did not occur because of pain or constriction of breathing (p. 873, Philinus' wife and Epicrates; p. 878, the Man of Clazomenae). But where pain is present and infrequent breathing with delirium is described, the pain must have been only heartburn (pp. 878–879, Dromeades' wife). Where a condition is described and a kind of breathing abnormality is mentioned that does not fit Galen's scheme, he explains that two pathological conditions must have been present and that the second, unmentioned, caused the abnormality (pp. 949–950, Nicostratus' wife, *Epidemics* 4; L 5.152).

Galen's argument is impromptu and poorly structured: he is clearly working out its shape as he goes along. Book 2 is devoted entirely to showing that Hippocrates did observe deep and infrequent respiration in delirium in cases reported in *Epidemics* 1 and 3 and interpreted in the sentence in *Prognostic* (ch. 5), "Deep and slow respiration indicates delirium." Book 3 attempts to prove that Hippocrates spoke of all of Galen's four categories of abnormality, plus the fifth (arrhythmic or unequal, more exhaled than inhaled or vice versa), and also showed explicitly or implicitly his awareness of their significance. In the process of his argument Galen is led to pronounce on the nature and the genuineness of various Hippocratic writings. He follows authorities (unnamed) "who seem to know the force of his works most accurately" (p. 855), who attribute three books of *Epidemics* (*Epidemics* 2, 4, 6) to Thessalus, Hippocrates' son; a combination of his own work and his father's notes, "found on skins and tablets" (p. 890). *Epidemics* 1 and 3 "are agreed" to be the only ones prepared by Hippocrates for publication, while "no one would think" *Epidemics* 5 and 7 to be worthy of Hippocrates. *Epidemics* 5 was attributed to Hippocrates' grandson, Hippocrates, son of Draco (p. 854). Galen's own opinion is that *Epidemics* 4 is not by Hippocrates or Thessalus (p. 891). *Prognostic, Aphorisms,* and *Regimen in Acute Diseases* "are reasonably believed to be by Hippocrates" (p. 891). Of the last named he says, "While out of jealousy they cheat Hippocrates out of many books, as not genuine, no one has dared to take this book away from the man. Many take away the part at the end, after the use of baths, but all preserve what precedes" (p. 913. One passage from the "spurious" part of the work [L. 2.414] is used in Galen's proof, p. 921, without comment as to its location in the work.). In a lengthy comparison of *Prognostic* with *Aphorisms,* Galen justifies *Prognostic*'s ignoring many matters because of its limited subject, acute diseases (pp. 930–938).

As his argument finally develops, Galen does not need fine

distinctions among Hippocratic works. He treats every mention
of abnormal breathing in *Epidemics* 1–6 and argues that each,
properly understood, is consistent with his theory (pp. 939–959).
Further, he promises that if he has the leisure he will do the
same for other works in the Corpus Hippocraticum, books (unspecified) by Thessalus, Polybus, and Euryphon (pp. 959–960).
At the beginning of his work he had proposed to show that
Hippocrates "far surpasses others in this subject, too" (p. 764).
Hence the rather inconsistent development of the work: Hippocrates, because of his peculiar greatness and brevity of expression, needs peculiar principles for interpretation. But once the
principles are established they can be applied to any work with
the same results. Galen's relationship to the authorities on Hippocratic texts is also ambivalent, depending on the requirements
of his argument: they "know best" (p. 890) and "know the force
of his works most accurately" (p. 855) when the categories of the
Epidemics are in question, but their jealousy (*philoneikia*) causes
them to cheat him out of other works (p. 913). He maintains a
tantalizing silence about who his authorities are, as he does about
the names of commentators, who "talk long and foolishly and
say nothing to the point" (p. 903). In the one question of a
reading in the text (in *Epidemics* 2, L 5.108), he speaks of "some"
who misunderstand and change it (p. 900). Later, in his commentary on *Epidemics* 2, he tells us that the "some" are Artemidorus Capiton and his followers (CMG 5.10.1, pp. 274–
275). Otherwise we cannot test precisely the depth or antiquity
of the judgments about Hippocratic works that Galen accepts
and disputes in this work. I will resume the question below, and
relate the canon of genuine works to the edition of Dioscurides.

   To sum up Galen's treatment of Hippocrates until the beginning of his commentaries on Hippocratic works: his repeated
homage to Hippocrates as guide and master is supported by very
little of substance beyond the brief memorized passages that
express the essence of a subject. Quotations from *Epidemics,
Aphorisms,* and *Prognostic* predominate, but many works provide
handy tags to illustrate aspects of Galen's doctrines. Galen never
enters serious discussion with anyone who has presented different views of Hippocrates, nor does he evidence study of

scholarly work on Hippocratic texts. His opponents are imaginary enemies of Hippocrates (as Praxagoras for his theory of the nerves and Theon for ignoring Hippocrates) or real but vulgar contenders in public brawls, whom Galen puts in the position of insulting himself and Hippocrates. The Methodist Thessalus' vague rejection of Hippocrates' harmful heritage in the introduction to his *Method* is the most substantial item for contention. Galen's relation to scholarship on Hippocratic texts and their transmission is largely antipathetic, nor does he consider himself in that scholarly tradition: they "cheat" Hippocrates of works, and so on. "Hippocrateans," to whom he refers a few times, are mentioned for their shortcomings: no one has yet really understood Hippocratic science, says Galen. But there is no serious discussion with Hippocrateans either: the few mentions seem adequately accounted for as references to his teachers, especially to the tradition of Sabinus.

With *Abnormal Breathing,* Galen enters a new phase in relation to Hippocratic texts. From that systematic commentary on one aspect of Hippocratic "doctrine" he soon proceeded, "at his friends' urgings" to commentary on the texts, at first with no more contact with scholarship or others' views than before, then in the later commentaries with attention to variant readings, old manuscripts, and other commentators. Never systematic in his scholarship, and apparently suppressing contrary opinions that weakened his own position, he nevertheless provided much material about the Hippocratic tradition. He had his own position to defend by the time he began his commentaries and would not change his views because of what he was now to learn.

GALEN'S HIPPOCRATIC SCHOLARSHIP

Dates of Galen's Commentaries and Related Works

ca. 175   *Abnormal Breathing*
Commentary on *Fractures*
Commentary on *Joints*
(Commentary on *Wounds,* now lost)

	(Commentary on *Wounds in the Head,* now lost)
	Commentary on *Aphorisms*
176–179	Commentary on *Epidemics* 1
	Commentary on *Prognostic*
	(Commentary on *Humors,* now lost)
	Commentary on *Regimen in Acute Diseases*
	Against Lycus
	Commentary on *The Surgery*
	Commentary on *Epidemics* 2
ca. 180	Commentary on *Prorrhetic*
ca. 186	Commentary on *Epidemics* 3
ca. 189	Commentary on *Epidemics* 6
	(Commentary on *Nutriment,* now lost)
	Commentary on *Nature of Man*
	(Commentary on *Airs Waters Places,* note 83 below)
	Hippocratic Glossary
	On the Order of His Own Books
	On His Own Books

Heretofore I have shown how Galen developed his original and comprehensive system, using and correcting predecessors and contemporaries, claiming himself the peculiar heir of Hippocratic science but hardly attempting to justify his claim, or, perhaps more accurately, taking his heritage as a spiritual one, inspiration rather than detailed science.

In the Hippocratic commentaries, which he wrote in his full maturity, one would hope for a breadth and comprehensiveness that would lead him to treat the Hippocratic tradition in historical perspective in relation to himself. One would hope, too, for evaluation of aberrations and successes of predecessors who had dealt with the Hippocratic Corpus. But Galen defines his purpose as commentator quite narrowly at each stage in his career. Though his definition changes slightly, he systematically deprecates "history" as antithetical to "science." He refuses to discuss anyone else's views at length, and when he accedes to friends' requests to discuss others' views, he tends to do so by picking out absurdities in their work. In the two commentaries where authorship and traditional views about authorship are most significant (*Prorrhetic* and *Nature of Man*), Galen literally suppresses information about other peoples' views. Nevertheless he gives us, almost by inadvertence, information that permits inferences about the material he is using at various stages and about the extent and depth of the tradition on which he draws. From that we can draw inferences about what else existed. The bulk of the writings makes them difficult to get at, so I shall proceed on the assumption that some characterization of the individual works will not be unwelcome to the reader.

He began his series of commentaries with the surgical works, the most literate, precise, and uncontroversial writings of the Hippocratic Corpus, which had set a standard in medicine for more than six centuries. In the first commentary he wrote, on *Fractures* (K 18 B.318–628), he set forth his principles as a commentator. The main and only essential purpose is to make clear what is unclear in the text (K 18 B.318). There are two kinds of unclarity; a communication can be unclear in itself or it can be unclear because the reader does not have the capacity or the background to understand it (p. 318).[49] In his exegesis Galen will, he says, write neither for the complete uninitiate nor for the excellently prepared reader, but for someone between the two (p. 321). It is not a part of exegesis to prove things true or false, nor to defend the text against sophists' interpretations, though that has become traditional in commentary and is acceptable (p. 318). But to argue about the author's views in definitive manner (*teleôs*) is not a part of exegesis, and he intends to eschew it (p. 319).

His method of exegesis, in the commentary on *Fractures,* follows his stated principles. He gives extended paraphrases of any statements that might be obscure and gives anatomical and physiological information that makes the procedures in *Fractures* reasonable. He notes one place where a "grammarian" altered the text mistakenly (p. 343) and speaks of the view of "some exegetes" on a passage (p. 418). In announcing that he will ne-

---

49. Galen says, p. 319, that he has written a brief work *On Exegesis* expanding his discussion of kinds of unclarity. I know nowhere else that he refers to the work.

glect all questions of orthography, he notes that "some" say Hippocrates' dialect is Old Attic (p. 322).[50] His only extended discussion of predecessors' views of the work concerns the title *Fractures* and the relation of the work to *Joints*. The first sentence of *Fractures* says, "The physician should make extensions as straight as possible in dislocations and fractures." This raises the question, why mention dislocations? Some people, says Galen, have asserted that *Fractures* and *Joints* are closely related parts of one large work which was entitled *Surgery*. They say that the small work now entitled *Surgery* was the work of Hippocrates, Gnosidicus' son (the Great Hippocrates' grandfather). The logic seems to be that such a minor work could not be by the Great Hippocrates. For convenience the great work was divided into two, and hence the titles of the resulting works do not quite fit—*Fractures* discusses dislocations in part, and *Joints* discusses fractures in part (pp. 323-324). Whereas others say that the two were originally separate works, titled by their primary contents, these people adduce the titles of other Hippocratic works to prove that such is his habit (p. 324). Galen himself has no opinion because the discussion is irrelevant to Science (p. 325). The question of his source is unanswered.

In the commentary on *Joints* (K 18 A.300–767), he repeats the discussion of the titles of *Fractures* and *Joints* in slightly different words, but with no more specific information (pp. 300–302). The commentary on *Joints* proceeds in the same manner as that on *Fractures*, with a few broad but vague gestures toward acknowledging his predecessors. He names "my teacher, Pelops" as the source of an interpretation of a term (p. 541) and speaks ironically of "some clever exegetes," who interpreted Hippocrates' use of superlatives (p. 663). Hippocrates says that the

50. Here Galen says that he has given his opinion about Hippocrates' dialect elsewhere, without saying where or what the opinion is. In later commentaries he says that Dioscurides and Capiton gave all Hippocrates' works in Coan dialect (*Epidemics* 6, CMG 5.10.2.2, p. 483); he also speaks of an ellipse as in the manner of the Ionians (*Regimen in Acute Diseases*, CMG 5.9.1, pp. 259–260); elsewhere he argues that the Ionians (therefore Hippocrates) wrote *kirsos*, not *krisos* (*Epidemics* 2, CMG 5.10.1, p. 175). On this subject, cf. Iwan Müller, Preface to *Scr. Min.* 2, xiii–xv.

shoulder is only dislocated into the armpit, while "others" disagree. Galen has seen five instances of outward dislocations, but
explains that he himself practiced in larger population centers
than Hippocrates (pp. 346–348). Promises by the author of
Joints to write other works are received casually by Galen. To the
promise to write a work on the nature of the glands, Galen says
that no such work by Hippocrates is preserved. The work by that
title in the Hippocratic Corpus was written by "some more recent Hippocratean," since it does not resemble the diction and
outlook of Hippocrates' genuine work. And he adds a tantalizing
statement: no previous (presumably previous to the *pinakes*)
physician mentions it, and the authors of the *pinakes* do not
know it (K 18 A.739).[51] Galen perhaps knows what he is talking
about, but those *pinakes* (presumably a catalogue of Hippocratic
works) are nowhere mentioned again, by Galen or anyone else. I
will suggest below what may be involved, when I discuss the
editions of Dioscurides and Capiton. When the author of *Joints*
promises to discuss further a subject in *Chronic Diseases of the
Lungs*, Galen remarks (p. 512) that many such promises in *Joints*
are not fulfilled—the promised work may have been written and
lost or never written. He adds that chronic lung affections are
discussed in the *Greater Book on Affections* (L 7.166–303) and in
the *First Book on Diseases* (our *Diseases* I, L 6.140–205), "wrongly
so titled" (and he gives the opening sentence of each book to
make the identification precise), but neither of these discusses
what Hippocrates promises in this passage. One presumes from
this that Galen is not claiming genuineness for the books he
names, but raising the subject without prejudice.[52]

The longest discussion in the commentary on *Joints* of previous arguments about the text relates to the setting of a dislocation of the thigh joint. As we shall see, the dispute had been a
commonplace for centuries. Galen says, "Some criticize Hippocrates" for setting the joint since it slips out immediately. Ctesias

51. Galen could not have accepted the work *Glands* as Hippocratic because it
conceives of the brain as a large sponge, not as the source of the nerves.
52. Littré records the various titles under which Galen cites *Internal Affections* in
different works (1.358–359).

of Cnidus was the first—"a relative of Hippocrates, since he was an Asclepiad by family"—and there were others after him (p. 731). Galen argues that the thigh joint can be successfully set and that he himself has treated two such cases (pp. 732–735; they require drying drugs and regimen since healing tendons are in question). He then cites Heraclides of Tarentum, "a man who did not lie for the sake of his dogma, as many dogmatics did," and presents a quotation from book 4 of Heraclides' *External Remedies,* which reads, in part: "Those who think the femur does not stay when it is set because of the tearing of the tendon that holds it in the socket of the hip are ignorant when they make a general denial. Hippocrates and Diocles would not have described how to set it, nor Phylotimus, Evenor, Neilius, Molpis, Nymphadorus, and others" (pp. 735–736). And Heraclides adds that he has succeeded in the operation on two boys.

For a perspective on this comment by Galen, let me quote part of what Celsus says on the subject (8.20.4): "Some say it always slips back out. But Hippocrates and Diocles and Phylotimus and Neilius and Heraclides of Tarentum, very well known authorities, say they have restored such cases totally."[53] Galen's comment and quotation from Heraclides reveal something of what is behind Celsus' statement. Galen's comment seems to be his own, and he livens the subject rhetorically by treating people's reservations about the operation as attacks on Hippocrates and by saying that those who deny its efficacy are liars in the service of their creed. The source of the strange statement that Ctesias (fifth century B.C.) criticized Hippocrates on the subject and that, as an Asclepiad, he was Hippocrates' relative, is difficult to imagine. But it is like Galen to populate history with quarrels against Hippocrates. Scholars have taken Galen's statement seriously—brief but precious evidence of the early knowledge of the work *Joints* (cf. L 1.70, 334), which they found important for the reconstruction of the Hippocratic tradition. Others have been more properly skeptical.[54]

---

53. *Auli Cornelii Celsi quae supersunt,* ed. Friedrich Marx (Leipzig, 1915), 8.20.4. See p. 213 below for Apollonius of Citium's presentation of another version of the same argument.
54. See, for example, Ludwig Edelstein, *RE* Suppl. 6, 1308 ("Hippocrates, Nachträge") with bibliography.

Finally, for Galen's procedure in these two early commentaries, two remarks are important: one, on the statement in *Joints* that the physician should know the nature of the spine and about the shamelessness of the empirics who called Hippocrates an Empiric (K 18 A.524–525), the other, the remark that people should pay attention to a particular passage if they think phlebotomy is to be used only in the plethoric syndrome, and especially Menodotus the Empiric (p. 575). These remarks, along with those mentioned above, confirm Galen's later description of his own early commentaries: he was simply giving his own view of the text as he proceeded, writing more or less in haste, and bringing up other people's views if he remembered them, but without close reference to anyone's work.

After the surgical works, Galen proceeds to commentary on the *Aphorisms*. The *Aphorisms*, a collection of descriptions of diseases and therapeutic measures, is only one of the representatives of that genre of medical compendium in the Corpus Hippocraticum, but it was generally considered the best of the aphoristic works. Galen attributes the excellence of *Aphorisms* to its organization on dogmatic scientific principles into which he has insight.[55] That argument is most thoroughly laid out in the commentary on book 3 which contains aphorisms on seasons and ages of man that Galen relates at length to his theory of

55. Galen makes an exception of *Aphorisms* 6.31 (K 18 A.45–50), which speaks of remedies for eye pains: Hippocrates must have arrived at these remedies by experience. He wrote them down "without stating the conditions logically or the syndromes empirically." Galen's teachers ignored the aphorism's teaching, but Galen, believing that Hippocrates would not have written it if he had not seen what he describes, worked out the distinctions and the proper conditions for the use of the remedies, neat wine (to drink), baths, vapor baths, bleeding, purging (he had told of this research also in *Therapeutic Method*, K 10.171–172) and found them superior to the drugs then in use. Galen's description of the situation seems to be confirmed by the earlier statements of Celsus, who quotes the aphorism as what Hippocrates says of eye treatment and notes that it does not say when or how to use the remedies, which is important (6.1.3). Celsus then gives recipes for drugs similar to those Galen claims to be replacing by his return to Hippocrates. It would seem, then, that Galen does respond to and correct a traditional interpretation of the aphorism. A later aphorism on eye pains (*Aph.* 6.46, K 18 A.151) Galen rejects as spurious: it has similarities with the other, but disagrees with it. He reports but rejects "commentators'" attempts to explain it.

temperaments (*kraseis*). He tells us that Lycus, "writing, as he says, the exegeses of Quintus, his teacher," failed to give a credible demonstration of what the aphorisms in that section said, but referred them all to experience and observation, although he did investigate logically the truth in other aphorisms (K 17 B.461–562). Lycus is Galen's only specific opponent in the commentary on *Aphorisms*, and even he is not attacked with the bitterness Galen later showed. Galen is also gentle in his general references to the Empirics, who took *krisis* in the first aphorism (experience is deceptive, judgment [*krisis*] is difficult) to mean judgment of medical remedies which experience discovers (K 17 B.354). Galen wants the contrast *peira* and *krisis* to mean experience and reasoning (*logos*) and thus to express the tenets of the logical sect. Much more frequently than in the commentaries on *Fractures* and *Joints*, Galen reports variant readings, and he frequently appeals to the reading shown by "most manuscripts" or "most commentators" (for example, K 17 B.727, 797, 825; 18 A.111, 113, 177). He reports one reading that was accepted by "the first commentators on the *Aphorisms*, among whom are the Herophilean Bacchius, and Heraclides and Zeuxis the Empirics" (K 18 A.186, on *Aph.* 7.69. This is one of the "spurious" aphorisms.). He reports two textual changes and one interpretation by Marinus (all in the seventh section, K 18 A.101–102). Galen says that most commentators are long-winded early in the commentaries but grow weary and gullible at the end. They fail to note that many aphorisms at the end are manifestly erroneous and not written by Hippocrates.

Whence came Galen's information for these reports about commentators and manuscripts?[56] Numerous as they are, I am inclined to attribute them to Galen's own notes accumulated mostly in his school days rather than to research concurrent with writing the commentary. He wrote his general report and criticism of Lycus' views without using or having at hand Lycus'

---

56. A naive treatment of the subject, which assumes that Galen ransacked bookstores and private and public libraries in the preparation of his commentary, is given by L. O. Broecker, "Die Methoden Galens in der literarischen Kritik," *Rheinisches Museum* 40 (1885), 415–438.

commentary. He saw Lycus' commentary, or more precisely, a section of it, only some time after he had finished his own, as he relates in a note added to his commentary at a later time. At the end of his explanation of *Aph.* 1.14, he adds:

> This will suffice for those who are inclined by nature to under-
> stand and trust what Hippocrates wrote. But against those who
> unfairly criticized him, like Lycus, I wrote a whole book answer-
> ing their attacks on this aphorism. The book is called *Against
> Lycus, that there is no error in the aphorism which begins "Growing things
> have most innate heat."* I was given Lycus' book after I had written
> my own commentary. Whence I added this statement, which was
> not in the earlier publication. I have made my defense against his
> criticism of Hippocrates separately in another book. [K 17
> B.414-415]

Even the work *Against Lycus* was written after Galen read only a portion of Lycus' work (CMG 5.10.3, p. 4). Hence Galen's remarks in this commentary about Lycus' versions of Quintus' interpretations would appear to come from hearsay. Similarly, Galen wrote his work on *Regimen in Acute Diseases according to Hippocrates* to counter an opinion he was told was contained in a book he had not read. Difficult for the fastidious scholar to imagine, the situation adds to our difficulty in assessing histori-cal information given by Galen. Galen had studied and thought about the *Aphorisms* for more than twenty-five years. His com-mentary fills out the results of his thoughts about "his" Hippoc-rates with material out of his marginalia and prodigious mem-ory, but without scholarly exactness.

These early Hippocratic commentaries are discussed in his later work *On His Own Books:*

> I knew that many before me had written comments on each
> phrase of his. But if they seemed wrong I thought it superfluous
> to refute them. And I showed that in what I first published for
> those who requested the commentaries, since I seldom said any-
> thing against those who had commented on them. For, in the first
> place, I did not have their commentaries in Rome, because all my
> books had stayed in Asia. If I remembered any exceedingly poor

statement by any of them which, if believed, would be greatly harmful to the work of the Science, I pointed that out, but all the rest I wrote according to my own view without mention of those who had given different explanations.   [*On His Own Books, Scr. Min.* 2.111–112]

He writes the following about reading his teachers' books when he was a student:

> If I find the meaning of individual expressions clear, and also correct, I do not refer to commentators on the Hippocratic works, so as not to be tedious. When I cannot get the sense of obscure passages, and there is no surety that I have understood the author's meaning correctly, I have introduced the commentators on this work whose commentaries are famous. I am aware of some of my teachers who invented explanations that their predecessors did not know. Among them are some who generally wrote nothing down, as Stratonicus, a student of Sabinus from my own city, and also Epicurus, an Empiric author also from my own city. In their books these people explained the writings of Hippocrates, but during their lifetime they published nothing. Only after their deaths the commentaries on some works appeared, and no small remainder stayed in the hands of people who did not publish them. Similarly, Pelops wrote commentaries on all the works of Hippocrates, but only a small part were preserved in public hands. Also my teacher Satyrus, and Philip, an Empiric, and other respectable men who lived in the time of my father and grandfather wrote many commentaries. I read most of them and made extracts from their writings. But I did not consider it good to introduce all these comments now into my commentaries, but limited myself to the famous, and to those who have offered something satisfactory in the interpretation of obscure passages.   [CMG 5.10.2.2, pp. 412–413]

In this last passage (from the commentary to *Epidemics* 6), Galen is explaining his procedure in that work and the sources for his repeated locutions, "some say," and the like. His early extracts from his teachers' books, initially selected for their medical usefulness, presumably served other purposes when he decided to write commentaries. They supplied the leaven in the commentary, the sense of sophistication in a tradition, which was achieved without actual active investigation of others' writings. I stress this because some modern scholars have assumed that Galen wrote with constant reference to previous commentaries or that he followed the commentary of someone who had, such as Sabinus.

The commentary on *Epidemics* 1, unlike that on *Aphorisms*, is virtually free of reflections of the centuries-long traditions of interpretation. Galen's introduction defines epidemic diseases, which he says is the work's subject, and summarizes his theory of *kraseis*, which explains their origin and makes prediction and treatment possible (CMG 5.10.1, pp. 3–6). He then discusses the epidemic constitutions and the case histories of *Epidemics* 1 as examples of the theory. Only Quintus is cited for an alternative view of the work: Quintus interpreted the third section of *Aphorisms* and *Epidemics* 1 badly, saying that Hippocrates' views came from experience, not from reasoning about causes (pp. 6, 52).[57] Galen also cites one change in the text by "those about Capiton" (p. 78).

The commentary on *Prognostic*, again, consists primarily of paraphrase, with frequent reference to the Science of the subject Galen had laid out in his own works *Crises* and *Critical Days*, written shortly before (cf. CMG 5.9.2, p. XVIII). Galen notices five variant readings from the editions of Dioscurides and Capiton, and he mentions various commentators' proposals as to what the "something divine" is that the Hippocratic text suggests the physician should look for (CMG 5.9.2, pp. 206–208. Galen concludes that it refers only to the condition of the atmosphere, a view he implies is original with himself. The commentators' views which he cites had appeared in Erotian's dictionary.). But

57. This is apparently a rewriting of his statement in the *Aphorisms* about Lycus' commentary. Galen does not mention Lycus as his source of information about Quintus (who wrote nothing), but uses Quintus' aberration to provide a rhetorical introduction to his own disquisition on the Hippocratic science which *Epidemics* 1 exemplifies. His later mention of Quintus' view that the place (Thasos) is not significant in a description of an epidemic constitution may come from his notes of Satyrus' exposition of Quintus' views (cf. CMG 5.10.2.1, p. 59).

in his usual manner of finding a rhetorical opponent, Galen offers one startling bit of historical information. In his exposition of the meanings of *prognosis, prorrhesis,* and *pronoia* at the opening of the work, he observes that the followers of Herophilus, who distinguish *prorrhesis* from *prognosis* as indicating, respectively, more and less certainty, are talking nonsense (p. 203). Later (p. 205), to lead into the commentary proper, Galen says,

> Perhaps I should not have mentioned Herophilus earlier. It is better to tell the truth as quickly as possible for those interested in the Science, and not exercise them doubly: once teaching the nonsense of the sophists and again refuting it. I think it would be better if I proceed with what is useful in this commentary, and then assign to a period of greater leisure and another work the examination of Herophilus' criticisms of Hippocrates' *Prognostic.*

But still later, when discussing the subject of the divine, and refusing to discuss it further, Galen says (p. 207), "I do not, as I said earlier, intend to notice everyone's errors, only what is credible. That is why I forebore discussing Herophilus' vicious writings against Hippocrates' prognoses." The last three words, "against Hippocrates' prognoses," can be taken to be the title of a work, an important fact, if it is one. Galen's refusal to discuss it or to give information and his escalation from "followers of Herophilus" to the viciousness of Herophilus himself cast suspicion on the reality of the supposed work. Handbooks and encylopedia articles have in the past said that this passage is evidence that Herophilus wrote a commentary (the view comes from Littré, 1.83), but it is not that.[58] Whatever it is, it sounds like something rattling around in Galen's memory that serves him to liven a page or two, not something he has in hand or gets out of a commentary that is before him. Further, whatever it is relates to concerns and quarrels currently exercising Galen. Prediction was of great importance to the reputation Galen was trying to establish for himself at the time.

Galen's work *Prognosis to Epigenes,* written at about this same

58. Edelstein, *RE* Suppl. 6, 1309, gives bibliography on the dispute.

time and filled with autobiography and self-justification, gives some insight into the contemporary social and professional quarrels he refers to. "Galen the mystery worker and miracle monger," his jealous contemporaries called him (K 14.641), because of his great success in predicting outcomes in disease and his tendency to take over other peoples' cases and embarrass them. As he quickly made his way in the imperial society, he acquired the enmity of those who had preceded him, as he describes in his encounter with Martialius (to embarrass whom he afterward wrote *Against Erasistratus*). Galen had predicted the course of Eudemus' fever and his recovery, when others had given him up. Eudemus had proclaimed that the Pythian Apollo spoke through the mouth of Galen. But Martialius confronted Galen in the street and asked whether Galen had read *Prorrhetic* 2 of Hippocrates or was wholly ignorant of it. "I told him that I had read it and that it seemed to me those physicians were right who did not think it belonged to the genuine works of Hippocrates, and that they were justified in saying so. 'At least,' said he, 'you know quite well what is written in it. Personally I do not make prophecies like that'" (K 14.619-620).[59] Martialius was referring to *Prorrhetic* 2's criticism of those who use flamboyant prophecy to make an impression. For example, "A man seems moribund to the physician treating him and to everyone else, but another physician enters and says 'the man will not die, but will be blind'" (L 9.6). For Martialius to quote Hippocrates is rather like the Devil's quoting scripture, and Galen resented it. This and similar references by Galen to current quarrels over prediction provide a context for his rhetorical flourish about Herophilus' viciousness. That Galen did not simply make up out of his head the book by Herophilus is proved because Soranus also spoke of a work by Herophilus against Hippocrates' *Prognostic* (Caelius Aurelianus, *Morb. Chron.* 4.8).[60] I will discuss below what it may mean.

---

59. Translation by Arthur J. Brock, *Greek Medicine: Being Extracts Illustrative of Medical Writers from Hippocrates to Galen* (London and New York, 1929), p. 210.
60. Caelius Aurelianus, *On Acute Diseases and On Chronic Diseases*, ed. and trans. Israel E. Drabkin (Chicago, 1950), p. 113.

Imaginary bits of information about the early history of the Hippocratic texts, such as Herophilus' commentary, are precious to modern scholars because real ones are lacking. Had Galen sought for genuine information he might have found it. That is to say, it is likely that there was information available to him which we will not now know. Significant "information" he gives in his next commentary on *Regimen in Acute Diseases* is correctly marked by Galen as conjecture, but scholars have, without exception so far as I know, distorted his statement and asserted that *Regimen in Acute Diseases* existed in its present form in the time of Erasistratus. Galen says, "This book [the second part of *Regimen in Acute Diseases*], even if it is not Hippocrates' work, is certainly old, so that it would have been added to the genuine portion already in the time of Erasistratus. Hence it is surprising that Erasistratus dared to ridicule Apollonius and Dexippus about their wax cups" (CMG 5.9.1, p. 277). This charming non-sequitur by Galen comes in his commentary on a statement in the text that one should give a fever patient as much water and honey water as he wishes to drink. Galen apparently reasons that Erasistratus should have read this, should have realized that though spurious it is a good reflection of Hippocrates' view, and hence should not have ridiculed Dexippus and Apollonius for giving tiny measures of water to their fever patients. To support "Erasistratus should have read this," Galen says that the section is old, though spurious, and therefore must have been read in Erasistratus' time as in Galen's. The conjecture is what Galen does with his own ignorance, and there is nothing reprehensible in it. But scholars in need of facts have not treated it with care.[61]

Galen's discussion of this passage offers another interesting example of the way he turns contemporary medical quarrels into judgments of the morals and motives of historical figures. Galen's theme in the commentary is that Hippocrates was the very first to bring method and Science to the treatment of acute dis-

---

61. For a recent treatment of Galen's statement as proof of the early condition of *Regimen in Acute Diseases,* see Iain M. Lonie, "The Hippocratic Treatise *peri diaites oxeon,*" *Sudhoffs Archiv* 49 (1965), 50. I know of no one who does not so misinterpret it.

eases, that he has not been surpassed, and that he is not adequately appreciated.[62] The Methodists criticize Hippocrates for overfeeding the ill; the Erasistrateans criticize him for starving them. Galen infers this, he says, because by slandering Hippocrates' students, Dexippus and Apollonius, Erasistratus was by implication slandering the teacher (CMG 5.9.1, p. 145). Erasistratus slanders Hippocrates' students but is unable to exhibit any book of theirs, and he pays no attention to Hippocrates' explicit words (p. 256). I am sure this last statement means that Erasistratus did not discuss or refer to Hippocrates and that he did not refer to specific books by Apollonius and Dexippus, but reported that they "are said" to have so treated patients. Very fortunately, though the Greek is lost, Galen's work on *Regimen in Acute Diseases according to Hippocrates* has turned up in Arabic translation.[63] That book was written for Victor, for whom a Methodist had written, asserting that Hippocrates properly fed patients daily. In his refutation, Galen actually quotes Erasistratus. Malcolm Lyons translates from the Arabic (CMG Supp. Or. 2, p. 107):

> They say that Apollonius and Dexippus, pupils of Hippocrates, who studied under him, used to prepare wax measures, twelve of which held one-sixth of a ratl [the Greek is cotyl; there was apparently about 1/144 pt. to a wax ladle] and they would daily ladle out three of these measures full of water to their patients. For the rest of their regimen they employed very severe restrictions and did not allow their patients to get anything to eat at all. For they thought that whatever moisture the patient received served like fuel as material for his fever.

62. On the basis of a passage that mentions discrepant opinions among physicians as among diviners (L 2.240–244), Galen argues that Hippocrates envisions a general method of judging disagreement in science by *peira*, test and experience. But no one before himself, he says, has interpreted the passage properly (pp. 129–131). He repeats the argument in *Opinions of Hippocrates and Plato*, in a passage apparently added after the commentary was written (Müller, pp. 781–783).
63. Lyons, CMG, Suppl. Or. 2. The Greek work of the same title, previously published in CMG 5.9.1, is clearly not by Galen though it appears to be contemporaneous with him:*De diaeta Hippocratis in morbis acutis*, ed. Ioannes Westenberger (Berlin, 1914).

Apparently, Erasistratus really did refer to Dexippus and Apollonius as "students of Hippocrates," and that is important. But, despite his rhetorical efforts, Galen cannot bring them satisfactorily into relation with his own Hippocrates, nor with *Regimen in Acute Diseases.* Anomalously, what scholars have repeated over and over as a fact—that Galen knew that the two parts of the work were read together in the time of Erasistratus—is wrong: Galen knew nothing of it but, for his own purposes, conjectured what must have been.[64]

In this same commentary, Galen's handling of the *Cnidian Opinions* is considerably more problematical. *Regimen in Acute Diseases* opens with the statement that the authors of the so-called *Cnidian Opinions* wrote accurately about the patient's experience in disease, but left out much of what the physician should know without the patient's telling him (L 2.224). Galen comments that they not only did not leave out what the patient suffers, but "went further than they should in describing some things, as I shall show shortly" (CMG 5.9.1, p. 117). Shortly later he recurs to that promise and gives information about what the Cnidian writers said. Unfortunately, the text is garbled, and the end of the first sentence seems to have been lost:

> I said above, at the beginning, that those from Cnidus wrote . . . "seven diseases of the gallbladder," and in the same manner again later, "from the bladder, twelve diseases," and again still later "four diseases of the kidneys" and again after that "from the bladder, four kinds of strangury," then later "four kinds of tetanus," and after that again "four kinds of icterus," and still later "three kinds of icterus." The point is that they looked to the varieties of symptoms which change for many reasons, and failed to consider the specificity of the dispositions as did Hippocrates—who used for their discovery a method only by using which one can find the number of diseases. And this

64. I discuss Dexippus and Apollonius below, p. 180. Galen's inconsistent judgments of Erasistratus in this commentary are noteworthy; because Erasistratus varied the regimen according to the patient's condition, he is Hippocrates' follower (pp. 125–127). Elsewhere, he viciously slandered Hippocrates (via his students) and ignored his works (pp. 256, 277).

method is described by me at the beginning of *Therapeutic Method,* is spoken of briefly in *Elements according to Hippocrates,* and in summary account you have our treatise the *Differentiation of Diseases.* [Pp. 121/122]

The end of this passage, like so many of Galen's references to Hippocratic science, is elusive, a kind of mystification by self-reference, because he does not actually treat of or demonstrate Hippocrates' method of enumerating diseases in those works, as he claims to have done.[65] Nevertheless, in his illustrations here of what Hippocrates is criticizing, Galen appears to be quoting passages or chapter headings from the *Cnidian Opinions:* "Seven diseases of the gallbladder," for example. Did he have the work in hand or does he draw on someone else's commentary that so

65. In *Differentiation of Diseases* (K 6.832–880), Galen analyzes a numeration of diseases according to hot, cold, wet, dry, and combinations, but without reference to Hippocrates. In the passages of the other two works suggested for comparison by Georg Helmreich (CMG 5.9.1, p. 122), Galen discusses numeration of elements in *Elements according to Hippocrates* (chs. 3–5, K 1.427 ff.), and in *Therapeutic Method* ridicules the Methodists for inability to define disease as opposed to symptom, affection, and the like (K 10.67 ff.). In 10.115–117, Galen talks about proper numeration of diseases, by *dyscrasiae* and by lack of congruity of corpuscles and pores, and at 117 he promises to show the "road" (*hodos*) of Hippocrates as developed by Aristotle and Theophrastus, that is, distinctions between *homoiomere* and organs which the "half empiric" Erasistratus and Herophilus never understood. The numerology of diseases and *homoiomere* versus organs was actually developed by Athenaeus and the Pneumatics, who cited the authority only of Aristotle and Theophrastus (K 1.519–523). Galen's attempts to connect the doctrine with Hippocrates all seem to come down to a sleight of hand by reference to his own previous works. In the *Opinions of Hippocrates and Plato,* Galen says that "Hippocrates censures the Cnidian physicians for their ignorance of the genera and species of diseases, and he points out the divisions by which what seems one becomes many by being divided: not only diseases but everything else, a matter in which one can find most notable physicians erring even to the remedies they use" (Müller, p. 776). This is followed by a passage Galen added to *Opinions of Hippocrates and Plato* after he wrote the commentary on *Regimen in Acute Diseases,* since it mentions the commentary and summarizes its results: Hippocrates said that a logical method is necessary—the Science cannot rest on empirical test and reach any kind of certainty. But Hippocrates, the first discoverer of such things, expressed himself in so disordered (*ataktoteron*) a fashion that virtually all commentators had missed his point (Müller, pp. 777–784). In the added passage, Galen does not try to demonstrate Hippocrates' method of division nor his enumeration of diseases.

illustrated the text?[66] Either seems possible. He does not recur to
the subject until the end of the commentary on the part he
considers genuine, where he sums up his observations about the
scientific outlook of the work by explaining how *Regimen in Acute
Diseases* fulfills its claim, implicit in its criticism of the *Cnidian
Opinions,* to tell the things the physician should know without
being told by the patient (p. 208). He makes it clear that others
have essayed the task before him: "Those who think Hippocrates is a dogmatic" said that what was meant was the places
affected, the patient's disposition, and the causes, whereas
"those who think he is an Empiric" said he referred to the seasons,
the characteristics of the various ages, and the condition of the
atmosphere. These observations are all useful, says Galen, but
he has others to add. He lists some significant general statements
from *Regimen in Acute Diseases* that treat of regulating wet and
dry food according to the patient's condition, observations about
the state of digestion in relation to diet, and observations about
the patient's temperament (characterized by yellow bile or black
bile), about coction of diseases, and looking to the crisis in
therapy. "All these and similar things were left out by the Cnidian physicians, but the physician must know it without the patients' telling him" (pp. 268–270).

Galen is here using the *Cnidian Opinions* and its authors as
foils, following the lead of the author of *Regimen in Acute Diseases.* He follows other commentators who did the same, but he
makes original contributions. Can we sort out what he inherited
from what he contributed? If we trust Galen's statements, he,
Galen, is the first to read into Hippocrates his own science of
enumerating diseases (which is essentially a development of
Pneumatic science). Other novelties in his interpretation are not
so clear: he accepts the Empirics' interpretation, which inter-

66. Littré assumed that Galen had the *Cnidian Opinions* "sous les yeux" (L 1.8,
2.200, 7.307). Johannes Ilberg, "Die Aerzteschule von Knidos," *Berichte der
Verhandlungen der sächsische Akademie der Wissenschaften* 76, no. 3 (Leipzig,
1925), 4, assumed that Galen used a source that depended on someone else
in Alexandria who had read the work. Late nineteenth-century source criticism intervened between the two assumptions. Possible alternative
assumptions—that Galen made it all up, that his source did, or that Galen
misread his source—have not been explored, to my knowledge.

prets Hippocrates by Hippocrates by adducing ideas from *Airs Waters Places* and perhaps *Aphorisms* 3 (*Nature of Man* is also possible); and he accepts the Dogmatics' interpretation, which is harder to see as interpretation of Hippocrates by Hippocrates: "causes, affected places, dispositions" are all catchwords of Galen's medicine (the first two provided titles to various of his works). Of course, one can find vague hints of such things in the Corpus, as Galen does, but no work lays them out, unlike the Empirics' reading of Hippocrates. That is to say that the Dogmatic interpretation which Galen reports seems to be a synthetic one like his own, a notion of science made up and then read into Hippocratic texts. Galen's own original additions to the reading of the science (with the exception of his numerology) seem more penetrating and closer to the text than the generalizations he quotes from the Dogmatics.

We cannot be certain whether Galen ever read the *Cnidian Opinions* or whether he got his few quotations from it from the commentators who had preceded him. I shall digress briefly on this difficult matter.

Galen quotes *Cnidian Opinions* once elsewhere, in the *Epidemics* 6 commentary, for its use of the word *pemphix,* and there he reports that "they" attribute the work to Euryphon (CMG 5.10.2.2, p. 54). In his composition of that commentary, Galen certainly drew on other commentaries. The citation of *pemphix* may well have come to Galen or his source from a dictionary. The fact that various classical poets and prose authors are cited immediately before the *Cnidian Opinions* rouses the suspicion that Bacchius (third century B.C.) is the ultimate source, since he habitually quoted poets to illustrate Hippocratic usage, culling his quotes from Aristophanes of Byzantium's dictionary. One apparent further mention of *Cnidian Opinions* occurs in the Galenic corpus, in a work whose genuineness is suspect.[67] Chapter 10 of *The Best Sect* cites "the Cnidians" as an example of Logical (that is, Dogmatic) thinking: they induced coughing to evacuate pus from suppurating lungs (K 1.128-129). Littré

---

67. Galen says in *On His Own Books* (*Scr. Min.* 2.120) that he wrote such a work, but Iwan von Müller's arguments that this is not it are compelling; see above pp. 94-5 and note 24.

(7.308–309) considers the method an example of the crudeness and primitiveness of the Cnidian School, but the author of *The Best Sect* does not.

Thus, Galen's works contain either two or three reports of *Cnidian Opinions,* none of which sheds light on the others and all of which may have been and likely were drawn from different sources. Important as are the failings he sees in *Cnidian Opinions* for the sake of his rhetorical argument in his commentary on *Regimen in Acute Diseases,* Galen shows a distinct lack of curiosity about the work or its authorship, there and elsewhere. The chapter headings Galen quotes from the work make it sound similar to the therapeutic works of the Corpus entitled Diseases and Affections, most especially *Internal Affections* (for details, see L 7.305), but he either does not notice the coincidence or sees no need to comment on it. Galen himself places the works on Diseases and Affections in the same category as the *notha* of *Regimen in Acute Diseases,* which he describes as follows:

> What is added to this book, after the part on baths, one might reasonably suspect was written by Hippocrates as an outline for his own memory, and that someone found it in his house after his death and published it. As a composition it is not worthy of Hippocrates' force (*dynamis*), as neither are the works entitled *Diseases* and *Affections,* though there is much in them that is well written. I will define and distinguish that in my coming commentaries on them.   [CMG 5.9.1, pp. 197–198]

Galen clearly intended at that time not only to write about works entitled Diseases and Affections, but also planned commentaries on gynecological and obstetrical material in the Corpus, at least on *Diseases of Women, Nature of the Embryo,* and *Eighth Month Child,* which he calls "Hippocrates' works" when he makes that promise (CMG 5.10.1, p. 297). Galen clearly knew nothing of the modern scholarly myth about a Cnidian school with distinctive doctrines.[68] We can only wonder what he would have produced had he attempted a full and consistent account of the matter.

68. I have written in "Galen on Coans vs. Cnidians," *Bulletin of the History of Medicine* 47 (1973), 569–585, of the absurdities scholarship has fallen into by vaguely citing Galen as the authority for the existence of a Cnidian school

In the part of his commentary that covers the *notha,* the spurious portion after the baths, Galen takes various attitudes toward individual items in the text. His introduction says that "many" have reasonably suspected spuriousness because it is inferior in force of expression and accuracy of *theoremata,* presumably meaning "theory" but also "observations." Still it is like Hippocrates, so perhaps a student wrote it. But then one might suspect interpolation because some statements are unworthy, and those mostly at the end, as at the end of the *Aphorisms, Wounds in the Head,* and *Epidemics 2.* And, in his manner, Galen offers us a chart: there are four possibilities: that statements are worthy in diction and attitude, defective in one, or defective in the other, or defective in both (pp. 271–272). Everything in the work is without order and like memoranda, as anyone can see, but much is so good that some readers have believed also in the erroneous parts, especially if they had no critical knowledge of the subject. So, for example, the excellent statements about when not to purge, for whose proper distinctions Galen refers to his own works on the subject (pp. 356–360). The statement in the *notha* on phlebotomy is so good that Galen is surprised it is not in *Aphorisms* (p. 286), and Menodotus the Empiric could have learned much from it (p. 287). Erasistratus should have read another excellent passage (p. 277, discussed above). But another passage should have added a qualification that can be found in the genuine portion (p. 332). The *notha* tend to unfounded generalizations, unlike the *Epidemics* of Hippocrates (p. 233). At one place Galen alludes to the absurdities of commentators who have misread the text (p. 301), at another to "the

medically and scientifically opposed to the Coan school, taking "Euryphon's" *Cnidian Opinions* as the source of all so-called Cnidian works and taking "Hippocrates'" words in *Regimen in Acute Diseases* as criticism of the opposed school. Galen and his sources do not so treat it, nor does Galen consider the works on Affections and Diseases Cnidian in their origin or type of medicine. Yet he considers the authors of the *Cnidian Opinions* scientifically inferior to Hippocrates. Because Galen is susceptible to anachronism in his rhetorical presentation, it is not surprising that the failings he attributes to the authors of the *Cnidian Opinions* are similar to those he attributes to the Methodists, whom he attacks in *Therapeutic Method:* they do not know how to number diseases properly or to take specific account of the patient's temperament, and so on.

majority" who attach a statement to what precedes, while "a few" treat it as separate (p. 355). And more vaguely: no worth can come from imprecise statements, but among other errors is commentators' tendency to enjoy vague statements.

Galen's vagueness about his authorities and predecessors is particularly exasperating here because various citations of the *notha* as Hippocrates' exist, particularly by Soranus (in Caelius Aurelianus' translation), who often cites Hippocrates' therapy as a bad example (see below, p. 225). But Celsus (first century A.D.) also cites a view from the *notha,* with approval, as Hippocrates' view (4.23.3). When were the particular stylistic judgments that Galen depends on made and by whom? When did the *notha* become *notha,* and how did the superior scientific precision of Hippocrates become a dogma? These are significant questions for constructing a history of the Hippocratic tradition. But Galen's approach and allusions to predecessors are of little help: some, he says, have judged stupidly; the *notha* are old, but they lack Hippocrates' finished precision and Science, and so on. In this period of his commentaries, Galen's scholarship is shallow. But by comparing these judgments with those of his later works, which more specifically report the general tradition, we can hope to evaluate them better.

One final example of Galen's narrow focus is *chondros:* did Hippocrates know it or not? Galen is much exercised about the subject in the early part of his commentary and in the work he wrote for Victor, as discussed above. *Regimen* and *Affections* and their provenance figure importantly in the evidence he adduces to prove that Hippocrates did know *chondros,* as did the ancient poets. But *chondros* is also mentioned in the *notha* of *Regimen in Acute Diseases:* "boil *semidalis, cenchros* or *chondros* in milk" (L 2.502). In his comment, Galen compares the qualities of chondros with those of the other foods mentioned in the passage (it is thick of substance, good and strengthening, [p. 355]), but he totally ignores the implications of the discussion for his passionate dispute earlier in the commentary. (Did his Methodist opponent read the *notha*? Did he read it as a work by Hippocrates? Did others who discussed whether *chondros* was a postclassical import from Scythia [CMG Supp. Or. 2, p. 105] discuss the question in relation to Hippocratic works?) Galen's resonant silences seem to indicate that he responds vigorously and rhetorically to a challenge that is on his mind, but afterward forgets it. Capacious as his memory appears to be, it has compartments. I have not successfully plotted out what a day's work on a commentary was in pages for Galen, to coordinate his memory with his day's work.

At about the time he wrote the commentary on *Regimen in Acute Diseases,* one of Galen's friends forced on him the commentary on *Aphorisms* by Lycus, or at least that part of it that made sport of the aphorism "Growing things have more innate heat." Galen had worked at interpreting that aphorism with originality, departing from those who thought it literally true that young things were hotter (cf. *Temperaments,* K 2.583–598), and had concluded that the innate heat, the source of growth and development, was what the aphorism refers to. Innate heat is a different kind of heat which one does not feel, but knows logically. Galen's idea was a development of Stoic and Pneumatic views. Lycus was more literal-minded: "Heat is heat," he said, and the aphorism is nonsense. Galen responded by writing *Against Lycus,* in which he calls the roll of philosophers who had distinguished qualities of whatever sort. Lycus' particular interpretive aberration, whether one call it empiric or something else, had not been in Galen's ken when he wrote his *Aphorisms* commentary. Hence the note he later inserted in that commentary (K 17 B.414–415). At about the same time that he wrote *Against Lycus,* Galen announced a more scholarly approach in his commentary on *The Surgery.* I cannot say that the two events are related. It may be that Galen got access to his books from Pergamum at this time.

The commentary on *The Surgery* announces a new interest in old manuscripts and old readings of texts:

> Some have tried to discover very ancient books written three hundred years ago. . . . I set about learning all that from the first commentators, so that I could discover the genuine readings from the best and greatest number of sources. The results exceeded my expectation. I found virtually complete agreement among the copies of the text and the commentaries, and I came to be amazed

at the recklessness of those who wrote commentaries in recent
times ("yesterday and the day before"), or who made their own
editions of all Hippocrates' work, among whom are those about
Dioscurides and Artemidorus Capiton, who introduced many
novel readings. It seemed to me that my account of the commentaries would be verbose if I noticed all readings. I decided it was
better to write only the old ones, adding some few, if they are
small changes and if all the commentators agree. The commentators are four: two wrote commentaries on all Hippocrates'
work—Zeuxis and Heraclides—and on less than all, Bacchius and
Asclepiades, misguidedly. But enough of that. [K 18 B.630–632]

I am almost certain that despite these apparently large claims
to have done research, actually Asclepiades is his main or only
source for early material in this particular commentary. The
statement about Bacchius as commentator is erroneous; Bacchius
wrote only a glossary that covered *The Surgery*, not a commentary.
Galen makes the correct statement in his next commentary
(*Epidemics* 2, CMG 5.10.1, p. 230). Zeuxis is not referred to
again in the commentary on *The Surgery*, and Heraclides is
cited once only, together with Asclepiades (p. 715), whereas
Asclepiades gets several other notices as well (pp. 666, 805,
810). Galen also makes his usual statements that commentators have left things unexplained, have disagreed, and so forth
(K 18 A.743, 748, 898–901). Although Galen is not much more
specific about the history of the text in this commentary than in
earlier ones, he does give some explicit information that leads to
some few inferences.

Two remarks indicate that predecessors had made the same
conjectures about this work as about others that were considered
not quite good enough and finished enough to be by Hippocrates. One is a simple aside, "Whether this work be by Hippocrates, or Thessalus, it does not talk about all medicine" (p. 664);
the other, "It is clear from this remark that this book was written
in notes and published after its author's death" (pp. 875–876, on
a repetitious passage which Galen conjectures is marginal notes
for possible revision which were incorporated into the text by a
copyist. Cf. also pp. 879, 880.). Galen's introduction gives some
indication of his predecessors' attitudes toward the Corpus: the

work's first paragraph, he says, contains a proemium to the whole science, which some thought should be read as an introduction (*didaskalia*), and some later classed it among what they called *eisagôgai* (works for beginners, the same as the title of Pelops' only known book). After its general introduction, Galen adds, the book teaches what is most useful for beginners in medical science (p. 632).

Galen's discussion of the work's opening paragraph is longwinded and not very explicit, but it gives some information about the kinds of commentary his predecessors had offered. Here is a fairly literal translation of the opening paragraph Galen comments on:

> What is like or unlike, beginning from the greatest and easiest, from those thoroughly and generally known, what can be seen, touched and heard. What can be perceived by eye and touch and hearing and nose and tongue and intelligence (*gnomê*). What can be known by all the things with which we know.

Galen tells us that there had been discussion whether the paragraph was repetitious and what theory of knowledge it contained. Aephicianus, Galen's teacher, adapting earlier interpretations, had interpreted it according to the psychology of Simias the Stoic and concluded that it is not repetitious: the senses described are the patient's; in the penultimate sentence the mind (*gnomê*) is that of the physician reasoning on the patient's experience, while the preceding words, seeing, hearing, refer to the physician's own observation (pp. 650–655). Galen finds this plausible, though he does not entirely agree, and he passes over implausible interpretations (p. 650). Galen says that Hippocrates must have added the final sentence, "What can be known by all the things with which we know," to avoid criticisms by sophists, who would say that he had not named all the means of knowing. "Don't think such things were not discussed in Hippocrates' time, even if we are not told about it," he says, and for the full solution of the problem he refers his reader to his own book on the *koinos logos* (not extant), a work on the reasoning faculty (pp. 657–664).

I would conclude, then, that in his plunge into research Galen did not go very deep the first time. He is pleased to get back behind the editions of Dioscurides and Capiton to recover earlier readings of the text, but his discussion of the meaning of the text is based on the Hippocrateans who were his immediate predecessors, Aephicianus, and perhaps Pelops. What Asclepiades did in his commentary besides note readings is left in the dark by Galen, and we can only wonder why. Galen always exhibited profound irritation at Asclepiades' medical theories and at his criticism of Hippocrates (cf. K 11.163). The motto attributed to Asclepiades, *tuto, celeriter, jucunde,* appears to me to have been based on *The Surgery,* which says, "do it quickly, painlessly, readily, neatly" (ch. 7; L 3.290). But Galen does not mention the matter. Also, I miss in Galen's commentary any reference to his earlier discussion in the commentaries on *Fractures* and *Joints* about the original makeup and provenance of the surgical works, and about the attribution of this work to Hippocrates' grandfather, a silence that adds to our impression that Galen is generally inattentive to what is not immediately in focus. The only purpose he claims in this commentary is to explain what is useful in this obscure little book about the equipment of the physician's office and procedures in bandaging. He has not entirely forgotten his earlier writings, in which he used passages of this work as evidence of Hippocrates' philosophy and primacy (pp. 708, 803), but does not consider them here. He neither defends the inferences he had previously drawn, nor treats the history of interpretations of the work except in the allusions I have cited.

The commentary on *Epidemics* 2, next in order of composition, is very expansive by comparison and full of informative remarks about, and quotes from, past commentators. We can virtually see the change occurring within the commentary, which begins (CMG 5.10.1) with the *explication de texte* of his early manner, then buds forth with a long quotation from Heraclides of Tarentum (p. 210) on an ambiguous word which Galen thinks Heraclides misunderstood. Thence the commentary continues to open up. Galen is not fully communicative by any means. He thinks it "makes no difference" whether he cites those who have interpreted a passage contrary (in his opinion) to Hippocrates' view, by finding the same doctrine as in *Ancient Medicine* (p. 220), and he sees no point in checking manuscripts or commentators on passages that are nonsense (pp. 300, 302). But still there is a fresh aura of information. Galen's new approach is partly the result of the nature of the Hippocratic work, as comparison with his later commentaries will show. *Epidemics* 2 and 6 are miscellanies of descriptions, prescriptions, and aphorisms, many of them unintelligible. When he does not feel confident, even Galen will go to a commentary. Partly also, judging from what he later said, Galen's new style resulted from pressure from his audience. The information about the history of the tradition he thus affords is useful, while we will still wish that we had similar information about the richer *Aphorisms* or the more exotic *Nutriment*. I shall discuss Galen's objectives in this commentary as well as try to give a picture of the kind of information he offers. Inevitably I must discuss his sources.

Crucial to Galen's picture of Hippocratic science is the anatomy of veins in *Epidemics* 2, which is very obscure but is the only anatomy of blood vessels in the Corpus Hippocraticum which can be reconciled with the facts that dissection reveals. The existence of the anatomy proves, says Galen, that the work is genuinely by Hippocrates (p. 310). If Hippocrates had written it for publication, not simply as notes, he would have added the significant things he left out (p. 393). Also, some of the text may be missing, Galen says (p. 321). If one adds to this anatomy the statement in *Nutriment* that the liver is the source of the veins (L 9.110) and that in *Humors* that the stomach is to animals as the earth is to plants (L 5.490), one can see what Hippocrates' physiological doctrines were (p. 313). Following the anatomy of veins in *Epidemics* 2, there is a description of two pairs of nerves, also obscure but correct, Galen says. Besides explaining the passages, Galen speaks of commentators who are ignorant of anatomy, but who have pretended to say something about the text (p. 238). He names only Sabinus, who talked about the necessity for veins in the body and who said the description of the nerves was very clear, thus proving that he had never seen a dissection (p. 329). Galen wishes he could resurrect Sabinus and debate with him (p. 333). But the correctness of the anatomy had been shown by Marinus (whether in a commentary is unclear [pp. 312, 329]), though even Marinus was wrong on at least one point (p. 327). Unfortunately, Galen never mentions Heraclides of Tarentum's commentary on the anatomical section of the work, nor that of Rufus of Ephesus, but elsewhere in the commentary Galen gives us some sense of their style.

He reports that Heraclides claims to have found a reading in an old manuscript (p. 220), thereby explaining, I think, his own vague statements on the subject in the commentary on *The Surgery*. And he makes the correct statement about early works on the Corpus: "The ancient explicators of the expression of Hippocrates, such as Bacchius and Glaucias, or those who explained his writings, such as Zeuxis and Heraclides of Tarentum (both belong to the Empiric sect) and Heraclides of Erythrae and other Dogmatics, cannot write this passage otherwise than I have given it. The physicians after them changed these readings variously" (p. 230).[69] With these words, Galen introduces a long discussion of a description of the chronic problems of a woman who had borne twins (L 5.92). As Galen writes it, the last sentence of the description means "Her tail, *ourai,* looked toward the temple of Aphrodite, *Aphrodision.*" He quotes at length Heraclides' argument that the passage cannot be interpreted satisfactorily and needs emendation (Heraclides, too, had a tantalizing manner of saying "some say"). Some said it was a metaphor for sexual desire, some said protrusion of the uterus toward the genitals, neither of which Heraclides can accept. For completeness, Heraclides quoted Bacchius' second book on Hippocrates' *lexeis.* Bacchius thought *Aphrodision* meant sexual desire. But, quoting passages elsewhere in *Epidemics* that describe where the patient lives, Heraclides proposed that we read *thyrai* instead of *ourai,* changing omicron to theta, thereby producing the meaning: "Her door opened toward Aphrodite's temple." In this way, Heraclides reduced the meaning of the passage to a statement about where the patient lived.

69. Elsewhere he calls Heraclides the "first commentator" on the work and uses the absence of a passage from Heraclides' work as evidence that the passage is interpolated (p. 284).

Galen admires the proposal and adds that later commentators (unnamed) took *thyrai* to be metaphoric for genitals and changed *Aphrodision* to *Aphrodisia* (sexual intercourse). Some changed *ourai* to *kourai,* "daughters," and said it meant that her disease lasted until her daughters were of marriageable age. Finally Capiton changed it to *rooi,* fluxes, and changed "looked" to "harmed" (*eblepon* to *eblapton*); "The flow from her genitals was bad for sexual congress." Galen draws the moral: do not change readings unless you know what the old manuscripts said. But for our sake the story Galen tells gives more insight into the tradition of the texts than anything in his preceding commentaries.[70] This material was unknown to Littré and to most scholars before the publication of Pfaff's edition of *Epidemics* 2.

As I have said, the expansiveness of Galen about the preceding tradition generally occurs in comments on passages like the above that are irrelevant to his own science, but he is sometimes revealing when he dismisses as trivial what others have taken seriously. I do not wish to proliferate examples, but some will be useful and will serve to characterize also the later commentary on *Epidemics* 6, where Galen's procedures are similar. Galen speaks of naive physicians, "who call themselves students of Hippocrates" and believe everything they find in books that go under his name. They lead their students astray and do great damage; this concerns a passage that prescribes hot and cold water in summer, neat wine and hot gruel in winter, for an ulcerated throat (p. 375, L 5.134.9-11).[71]

On a passage which says that blood gathering in the breasts is a sign of incipient madness (L 5.138.19), Galen remarks that this is a symptom, but one that does not occur in all cases of madness. He adds that if the reader wishes to know what those say who

70. The discussion covers pp. 230-234 in Pfaff's CMG edition. I summarize it as I understand it, making some changes from Pfaff, who has, I assume, translated the Arabic without alteration. Littré's judgments on the text of the passage (5.92) are based on the forged Galenic commentary on *Epidemics* 2.
71. In his *Epidemics* 3 commentary, writing on gullible commentators, Galen says that the followers of Sabinus and Metrodorus (*hoi peri,* a locution by which Galen probably means the men themselves) claim to be more accurate than previous Hippocrateans, but in fact they make many errors (CMG 5.10.2.1, pp. 17-18).

look for a cause, he will tell him: Rufus says that it is only significant after puberty, and with a great quantity of blood, whose
strong heat supports the dryness that causes madness. Sabinus,
however, gives various contradictory explanations, which Galen
gleefully quotes: (1) poor nourishment; (2) thrust of blood
caused by upset nerves; (3) that 2 can't be true because the
nerves are not in the heart, but are where the seat of reason is;
(4) perhaps 2 is true because excited nerves press on breast and
then on heart, causing blood flow; (5) blood collecting is corrupted and turns to black bile, which causes melancholia. Galen
ridicules Sabinus' inconsistencies and errors (pp. 408–409).

On a passage which I translate here without punctuation because each commentator supplies his own, Galen gives a long,
rambling description of types of approach by commentators:
"Ruddy sharp-nosed large eyed base ruddy flat-nosed large eyed
good hydropics blue-eyed ruddy sharp-nosed unless bald" (L
5.128.1–3). Three types of countenance are described here, says
Galen: the first is of good character, the second is bad, the third
is not characterized, but the word "hydropics" between the second and third can apply to either of them. Commentators divide
into two general groups, but he names only Numesianus as
Pelops reported him (remember that Galen studied with both,
but read only Pelops' Hippocratic interpretations) and Heraclides. The latter interpreted the passage as a description of
character type as shown in physiognomy, the former as related
to hydropsy and exemplifying the dictum that disease which
expresses the body's nature is easy to heal, disease which opposes
it is difficult. But some say red means warmth, some say cold,
others say cold and moistness, and the other features are problematic in the same way. Galen does not have a solution to offer,
though he can find error and contradictions in all the discussions that he reports. His own view is that one cannot tell the
krasis of the whole body simply from physiognomy (pp. 345–
351).[72] Interestingly, Pelops cited *Epidemics* 5 and *On Vision (De*

72. It is interesting to compare Galen's certitude about the significance of a
nickname "Griffinfox" for the physiognomy of a patient in *Epidemics* 6 for
explanation of his symptoms: nocturnal emissions and *phthisis* (CMG
5.10.2.2, p. 503). Surely this comes from similar traditions of interpretation,

visu), among other Hippocratic texts, to prove his point, the only citation of the latter in all Galen's works (cf. Littré, 1.116, who thought the work was unknown entirely in antiquity).

What are Galen's sources in this commentary? Pelops' memory of the interpretation of Numesianus comes from Galen's early notes (see above, p. 69). His frequent references to Dioscurides and Capiton probably mean that he is using their edition. Beyond that, his citations of and quotations from Sabinus, Rufus, and Heraclides of Tarentum are a new scholarly element in this commentary, compared to his earlier ones. Is he using the work of one only (Sabinus) or of two or of all three?[73] There is no way to be certain, but the question is not crucial. In my view it is likely that all the old material, including the quotations from Heraclides, comes largely from the commentaries of Rufus of

---

though Galen, being certain of the truth in the case of Griffinfox, names no sources. In his *Hippocratic Glossary*, Galen cites Griffinfox (*grypalopex*) in a list of compound adjectives describing medical characteristics drawn from the Corpus, along with one drawn from Erasistratus (K 19.142). The list follows a citation of Dioscurides and would appear to be drawn from his *Hippocratic Dictionary*.

73. Max Wellmann, "Zur Geschichte der Medicin im Altertum," *Hermes* 47 (1912), 4–17, set the problem. He said that Galen did not use Rufus directly, but that Sabinus was the source (the two are generally cited by Galen one after the other or not at all). Some of Wellmann's evidence is invalid, including the forged commentaries on *Humors* and *Epidemics* 2. The credibility of his argument depends heavily on our belief that positing the minimum number of sources is the correct procedure and that Galen everywhere worked in the same way (even in his early commentaries, an assumption I have shown to be wrong). Wellmann's general conclusions have been accepted, although details have been disputed: see H. Diller, *Gnomon* 22 (1950), 231–232 and the literature cited there. I see no way to settle the question finally, but I reject the assumption of a single source and uniform method, which would have been too obvious to the literary circle Galen addressed. There is a question of how much pious fraud one can accuse Galen of when he says, on a passage in *Epidemics* 6, where he has quoted many variants, "I have to gather together the explanations of all the commentators on this passage.... Where I have a certain knowledge of what Hippocrates meant I consider it superfluous to read what Zeuxis and Heraclides of Tarentum, and those after them who have explained Hippocrates' works, say about the meaning of every expression of Hippocrates" (CMG 5.10.2.2, pp. 289–290). Further, if we are reducing sources by hypothesis, how do we take the statement, "Sabinus is the only one I found writing it so, and Metrodorus followed him, and those after him until now" (CMG 5.10.2.2, pp. 46–47)?

Ephesus. It seems barely possible that Galen knew no more about Rufus than what he derived from Sabinus' commentary. Galen's quotations of Rufus in his commentaries on *Epidemics* 2, 3, and 6 and on *Prorrhetic* indicate that Rufus habitually quotes the more ancient commentators and concerns himself with textual matters.[74] Galen's quotations of Sabinus do not give the same impression, nor, I think, is it anywhere demonstrable that Sabinus quoted or criticized Rufus. Galen cites Rufus but not Sabinus in the commentary on *Prorrhetic*, in connection with which we will consider further Galen's method of planning and writing his commentaries. First, we may notice that around this time Galen wrote a small work, *Seven Months Child*, which though its existence was known, was available only in Arabic before Richard Walzer offered a German translation in 1935. The work is an exercise by Galen in creating an imaginary historical situation in order to "save" Hippocratic works from "previous interpreters" (unnamed), who would deny them to Hippocrates because they are inconsistent with one another on a rather concrete point, the number of days from conception to birth of seventh-month babies.[75] Galen resolves the inconsistency with a conjecture (in which he implies that he is original) that two of the works in question, *Epidemics* 2 and *Nutriment*, were written in Hippocrates' youth, when he did not know that months cannot be reduced to whole days, a view Hippocrates later asserts in *Prognostic*, and that the third, *Eighth Month Child*, was written in his maturity (Walzer, p. 347). But, Galen claims, the numbers offered in all the works are within the realm of possibility.[76] In presenting the problem, Galen reports, with his usual reticence about who made what judgment, predecessors' conjectures

74. See, for example, CMG 5.10.2.2, pp. 174 and 411, and, on *Prorrhetic*, CMG 5.9.2, p. 73.
75. Richard Walzer, "Galen's Schrift 'Ueber die Siebenmonatskinder,'" *Revista degli Studi orientali* 15 (1935), 323–357. The work is to be dated sometime after the commentary on *Epidemics* 2 (see ibid., pp. 330–331).
76. Hippocrates' "most genuine works," *Aphorisms* and *Epidemics* 1 and 3, Galen argues, agree with *Prognostic* in reckoning neither years nor months by full days (Walzer, pp. 347–348), and he refers to his own exposition of the subject in *Critical Days*.

about the provenance of works of the Corpus, some of which he notices nowhere else: "some" say that all the works of the Corpus are by Hippocrates; "some" say that there were four people named Hippocrates: his grandfather, himself, and two grandsons, all of whom are represented by works of the Corpus. "Some" attribute the whole Corpus to those four, while "others" attribute *Eighth Month Child* to Polybus and *Nutriment* to Thessalus. "Some" attribute *Epidemics* 2 to Thessalus; "others" call it a forgery. "Some" say that *Nutriment* is by a Herophilean and some say by an unknown author. They point out that statements in *Epidemics* 2 and 6 are not genuine: those works are what Hippocrates left in notes on paper, skins, and tablets, which his son Thessalus assembled. But Thessalus made interpolations in the work, and "some" think that others did also (Walzer, pp. 345–346). Galen does not say that people actually had argued about genuineness on the basis of the discrepancy in numbers which he sets out to solve, nor does he explain the specific occasion for his writing the work. He seems to offer the plethora of possibilities of authorship as something of a *reductio,* as if to say, "see what confusion the interpreters have fallen into," before he offers his own solution of the numerical discrepancy.

Now we proceed to the commentary on *Prorrhetic,* which seems to have been written out of spite against Lycus, Galen's perennial bête noire. Just as Galen had been incensed when he was shown parts of Lycus' commentary on the *Aphorisms* after he had written his own, so he now encountered Lycus' writings on *Epidemics* 3, or at least on the first part, as he was about to write his own commentary on that work. His readers had requested that he take account of commentators with whom he disagreed. In his commentary on the first case history in *Epidemics* 3, Lycus, Galen tells us, adduced three items from *Prorrhetic* to explain the symptoms and pointed out that *Aphorisms* 4.49 was wrong in its prognosis of death. (Galen says that the items from *Prorrhetic* do not apply and that Lycus had misread the *Aphorisms*.) Lycus, "the bastard of the Hippocratic sect," did not talk of humors or *krasis* in his comment, nor mention the neglect of phlebotomy in the handling of the case. In short, he acted like an Empiric (CMG

5.10.2.1, pp. 13–17).[77] Hence Galen wrote the commentary on
Prorrhetic to "deprive" Hippocrates of the work, which he had
not found offensive or un-Hippocratic before. Indeed, he had
cited *Prorrhetic* to show that Hippocrates' view of coma accorded
with his own and had found in it the characteristic locution of
the scientific Hippocrates, *skepteon*, "one must investigate" (CMG
5.9.2, pp. 181, 190, 192). Nowhere in this commentary on
Prorrhetic does Galen mention the fact that he is writing it to spite
Lycus, nor does he mention his own former views (note CMG
5.9.2, pp. 6 and 29). He attacks *Prorrhetic* at length as an imprecise work, empirically oriented, which ignores causes and uni-
versals and proceeds in a way that would take the student a
thousand years to learn medicine (CMG 5.9.2, pp. 32–33).
Within the commentary he is elusive on the subject of everyone's
views about the authorship and nature of the work except his
own.[78] Here is Galen's own description of the commentary and
its place in his career, from his later commentary on *Epidemics* 3:

> Since I knew that I had always explained Hippocrates' view in all
> the works I had written, and quoted his timeliest remarks, I
> thought it superfluous to write exegesis in commentaries on each
> phrase from beginning to end of all his works. But since some of
> my friends begged to have these, too, I started in from the most
> genuine and useful of Hippocrates' books, keeping to the same
> outlook as in the treatises I had previously written: what is seen
> for long periods or frequently in illness is to be described, with-
> out refuting those who have written down the infrequent, the

77. In the interpretations of Lycus that Galen later quotes in the commentary
on *Epidemics* 6, however, Lycus talks about the humoral pathology which the
cryptic statements of that work refer to (CMG 5.10.2.2, pp. 286, 291). Galen
probably exaggerates considerably in his commentary on *Epidemics* 3.

78. I think that the editor, Hermann Diels, is mistaken in his assumption that
the abrupt beginning of the commentary indicates that a proeme has been
lost (CMG 5.9.2, p. ix). Diels's somewhat vague appeal to Galen's habits
elsewhere fails because Galen's habits elsewhere are very various. He begins
the *Prorrhetic* commentary as he began the one on *Prognostic*, discussing
prorrhesis and *prognosis* and rejecting Herophilus' dictum. Here, more accurately than there, I think, he does not speak of a Herophilean book against
Hippocrates' prognoses, nor does he promise to write against it (CMG 5.9.2,
p. 1).

never seen, and the false. And thus I proceeded, in the commentary on *Fractures* and *Joints,* then in those on *Ulcers* and *Wounds in the Head* and those on *Aphorisms* and *Prognostic.* And after that, obedient to my friends' urgings, I wrote the commentaries on *Regimen in Acute Diseases,* the first on the part of the book up to the uses of baths, which is believed to be the most genuine, and the second on what follows, since this also seems to have many observations that accord with Hippocrates' view. And after that I wrote the commentary on *Humors* quickly, in a few days, hurrying because of the impending journey of the man who asked me to write it. And since these commentaries were well thought of (they circulated widely), many, not only those of my circle but other physicians, urged me to write commentaries on all Hippocrates' work. So I did the *Surgery* and *Epidemics* 1 and 2, and was going to do *Epidemics* 3 when I turned aside to *Prorrhetic* at the strong urging of some friends. The urging resulted from a remark I made by the way regarding what is said in the *Prorrhetic* and *Aphorisms.* I showed that all their contents had great force in medicine, but that if one holds to all that is written in *Prorrhetic* as though it is generally applicable he will often err. And I showed that most of the things in the *Coan Prognoses* are of that sort, mixed in with things which are said in the *Aphorisms* or *Prognostic* or in the *Epidemics,* and that only its parts that are written in those books are true. But all the rest is bad in *Prorrhetic* and *Coan Prognoses.* . . .

When I had finished the comments on *Epidemics* 2, as I said, I wrote in the interval three comments on *Prorrhetic* because of my friends' insistence, and in them I showed that most of the things in the book are erroneous: many things being spoken of as universal, though in fact they are rare, and various things ill defined and lacking distinctions. For clarity's sake I shall recall one or two examples to illustrate the faulty explanations given by some in their commentaries on the *Prorrhetic* and *Coan Prognoses* and on the patients in the *Epidemics.* There is in *Prorrhetic* a statement to the effect that fevers in the hypochrondria are pernicious. Some of the commentators on the book remind us of patients in the *Epidemics,* citing only those who died, and of whom Hippocrates wrote that there was stretching of the hypochondria in relation to the occasion of the illness, pain, trembling, or the like. But the commentators are silent about the patients who recovered. Whereas Hippocrates, in the same way that he wrote the rest of the useful information about acute illnesses in *Prognostic,* also wrote about the hypochondria in that portion of the work that begins as follows: "It is best for the hypochondria to be free from pain, soft, and with the left and right sides even." This is the beginning of the whole lengthy discussion in which Hippocrates teaches about the hypochondria. But one who says simply that fevers of the hypochondria are pernicious without the appropriate definitions that Hippocrates wrote in the *Prognostic* is wrong. It would be much truer to say fevers in the head, if it is ill, are pernicious.   [CMG 5.10.2.1, pp. 60–64]

Galen's elusiveness within the commentary on *Prorrhetic* and his refusal to name his sources or give their views is particularly obvious within the context of his new researches. He repeatedly mentions commentators (unnamed) who, for example, adduce cases from the *Epidemics* to illustrate symptoms and prognoses in *Prorrhetic* (CMG 5.9.2, pp. 15–16). But they are foolish because they do not know the distinctions Galen made in *Affected Places, Abnormal Breathing*, or the work on *Tremor, Palpitation*, etc. (pp. 16, 19, 116). Everybody before him, he says, explained the writing mistakenly (p. 30). Yet he appeals to the opinion of "some" that the work is not Hippocratic because of its solecisms (pp. 13–14) and elsewhere says:

> I will say again what I said before. The writer of this book seems to be of the same science as Hippocrates, but far inferior to him. For that reason some have thought it was by Draco, Hippocrates' son, some by Thessalus. (It is agreed that there were two sons of the Great Hippocrates, Draco and Thessalus, and each again named a son Hippocrates.) Whether this book was written by one of them or by someone else, or whether the author died before he could publish it, is not worth pursuing.   [Pp. 67–68]

Obviously, *Prorrhetic* had been subject to the same stylistic conjectures as other works in the Corpus by Galen's predecessors. But clearly, also, no one before Galen had damned it on the grounds he does. He attributes to the work the sins he found in Lycus. I suspect that had he wished to prove the opposite, that Hippocratic science is clear in the work, he could have.

Besides readings in the editions of Dioscorides and Capiton, Galen mentions only an interpretation by Satyrus, a reading found in an old manuscript by Quintus (probably both from his old notes on Satyrus' teaching), and a criticism of Zeuxis by Rufus, which tells us about Galen's method of work:

> Most copies have *oura pepona,* and the commentators know it as so written, but Rufus of Ephesus, a man who always tries to preserve the ancient readings, here criticizes Zeuxis, the earliest Empiric to write a commentary on all Hippocrates' books, and says the following: "Zeuxis, if I ought to mention him, too, avoids silliness generally, but here shows it by falling into error and then defending it. He wants the text to read *oura pepona,* as much as to say 'urine with pus and thick matter is bad.' He is unaware that this is numbered among the greatest benefits." So wrote Rufus, not unworthily, against Zeuxis, saying the reading *pepona* was wrong, and preferring the reading *epipona,* meaning painful excretion. Those who accept the reading *pepona* say incredible things unworthy of mention, although there are creditable arguments they could have used.   [CMG 5.9.2, p. 73]

Here Galen accepts Rufus' worthy criticism of Zeuxis, but later in the treatise Galen encounters the expression again.

> Why pursue what he means by *oura pepona?* On that subject I said above that Rufus of Ephesus had criticized Zeuxis the Empiric, for the statement is false. Such [concocted] urine never causes trouble. But it is true that *epipona oura* indicates something unfavorable, namely voiding with pain. Still, this is the second occasion for writing *pepona oura,* and it is no longer credible that the text is faulty in both passages, as Rufus thought when he criticized Zeuxis.   [P. 121]

If we think about the papyrus rolls Galen worked with, it adds a new dimension to our problem of his sources. Galen went from one end of a roll to the other, working in more or less haste. He knew what he had written (often he mentions the possibility of marginal corrections or suggestions by the author himself as a source of confusion for copyists, and he attributes the same suspicions to the editor Dioscurides [CMG 5.10.2.2, p. 464]). He knew what he had read, but he could be surprised by what turned up. He could not thumb back and forth. The choice between saying, "commentators talk nonsense," since he is certain that they do, and unrolling their books one by one to check a passage, must have faced him often, whatever the size of his library. Galen planned in advance how many books (that is, rolls) he would fill with a work and governed his prolixity accordingly (as he states in his commentaries on *Epidemics* 3 and 6 [CMG 5.10.2.1, p. 86; 5.10.2.2, pp. 177, 218]). When we add his repeated assertions that his only interest is in the Science as he sees it and that he consults and reports predecessors only when pressed by readers or when the text is unintelligible to him, we should conclude that we cannot generalize about his methods of research, but must establish the likelihoods for each case in each commentary, within the limits of the material he apparently had at hand at the time. Hence, the generalizations by Wellmann and Broecker about Galen's methods should be discarded.

Before proceeding to *Epidemics* 3, the next commentary, we can usefully glance aside here at Galen's *Hippocratic Glossary* (*Linguarum seu dictionum exoletarum Hippocratica explicatio*, K 19.62–157).[79] The *Glossary* is a book clearly made from other people's books, whose sources might be many or very few. It does not, as do Galen's commentaries, address itself to the Science, but has antiquarian, literary, and linguistic purposes. The *Glossary* is a list of obsolete words and of words that are used in a peculiar manner in the Corpus, a special study which Galen says that he did for one Teuthras: Teuthras had the commentaries of Galen (it is unclear how many), but wanted a brief summary of the glosses with definitions (p. 65). The *Glossary* was definitely written after the little book *On Coma*, to which it refers (p. 65). Otherwise I find no clear indications of date. Galen may have used as his basis Dioscurides' explanation of Hippocrates' diction, which Galen cites repeatedly throughout the *Glossary* for definitions and textual variants, and simply culled his glosses

79. Not much has been written about Galen's glossary. Konrad Schubring before his death was preparing an edition of it for the CMG. Hans Diller successfully defended its authenticity in "Zur Hippokratesauffassung des Galenos," pp. 167–181. Johannes Ilberg offered some observations about its sources in "De Galeni vocum hippocraticorum glossario," *Commentationes Ribbeck* (Leipzig, 1888), pp. 327–354. A full study remains to be done.

from it, taking from it at the same time citations of authorities: Dioscurides of Anazarbus' *Materia Medica* (pp. 99, 105ff.), Theophrastus' *On Plants* (pp. 91, 109), Zenodotus' *On Dialectal Words* (p. 129), Menetheus' *Names of Drugs* (p. 89), Kritias' *On the Nature of Love and Virtue* (p. 94), along with citations of those who had written on Hippocratic works, the names of whose writings are not given: Erotian, Zeno the Herophilean, and Zeuxis (p. 107).

In his introduction, where he describes the nature and purpose of his work, Galen criticizes Dioscurides' dictionary, in an oddly churlish way: Dioscurides' work claims to be an explanation of all Hippocrates' diction, yet has barely a third or a fourth; it explains terms that are obvious and explains plants, animals, fish, geographical terms, stars, things even a child should know, drawing shamelessly from writers on those subjects (pp. 63–64). He, Galen, will do as Bacchius did, who wrote only on glosses and drew his examples from Aristophanes (of Byzantium, our Mss. say Aristarchus), but he will offer glosses that Bacchius left out and will omit no gloss (pp. 65, 68). He will draw them from the whole Corpus, not only from the genuine works (p. 68). To Dioscurides and others like him Galen refers those who want definitions of obvious words (p. 68).

Hence, Galen claims no originality save in the selection of words, and he gives no indication that he offers anything that was not in Dioscurides' dictionary. Where Galen overlaps what is extant from Erotian's dictionary (for example, in the definitions of *kammoron, amphidexios, iktar*), his definitions and citations of authorities seem to be part of the standard lexicographical tradition. But there is one striking novelty in Galen's *Glossary:* it is in absolute, or virtually absolute, alphabetical order—it is alphabetized like modern dictionaries and not only by the first or first two or three letters. It is the earliest known example of such alphabetization, but Galen does not mention this feature. He promises only that the "order will be the order of the letters that begin the words, as you bade me" (p. 62). To develop the technique and to expend the effort to put words in such order is no inconsiderable accomplishment, as Lloyd Daly has shown. Daly found no discussion or description of such alphabetization before the eleventh century and found Galen's use of it isolated and virtually unique.[80] Since we might expect some mention of the technique in his introduction had Galen himself made the advance (such as cutting a word list into strips and pasting them in order), we might conjecture that Dioscurides preceded him in so ordering the words.

To return now to the commentaries: the commentary on *Epidemics* 3 is the next one Galen wrote. *Epidemics* 3 resembles *Epidemics* 1, being comprised of individual case histories and histories of the year's weather and resulting diseases. His comments generally follow his earlier manner: fairly brief paraphrase with diagnosis and explanation in his own terms of the phenomena described. It is therefore a short commentary, which he intended from the beginning to execute in three books. Yet it retains some of the tone of the immediately preceding informative commentaries and gives evidence of his new stage in scholarship. He turns aside from his main business from time to time to abuse Sabinus' commentary for its irrelevance to the Science. In deference to readers' requests, so he tells us, he turns aside after discussing the first case history to discuss inferior types of commentaries. Unfortunately, his discussion is in such general terms as to be not very informative. The bad commentaries: (1) attribute significance to the patients' addresses in diagnosing their diseases; (2) find the patients' names significant or discuss matters of orthography; (3) are similar to those of Lycus in that they use *Coan Prognoses* and *Prorrhetic*, do not speak of humors in their diagnoses, and so on; and (4) are like those of Sabinus and his followers, who turn out to be an example of the first category: they thought that the patient of the first case history had an illness caused by sexual continence because the text says that he lived by the temple of Earth (CMG 5.10.2.1, pp. 11-25). Galen's only specific examples of bad commentators, therefore, appear to be Sabinus, Lycus, and an unnamed professor in Alexandria who made a joke on the name Silenus, the patient in *Epidemics* 1 whose symptoms were nocturnal restless-

80. Lloyd W. Daly, *Contributions to a History of Alphabetization in Antiquity and the Middle Ages,* Collection Latomus 90 (1967), pp. 34-35.

ness, talk, and laughter (p. 12). This gives an impression of a
very shallow tradition on which Galen draws. Had we only this
commentary, we might well follow Wellmann in inferring that
Sabinus was the sole source of Galen's historical information.

Galen's other reluctant historical digression in this work is of
much greater significance. The case histories in *Epidemics* 3 are
followed by shorthand symbols, Greek letters that seem to summarize in code the salient factors in the disease. Their meaning
and origin had been disputed by the earliest students of the text.
Galen would like to dismiss the subject, but feels that he cannot.
He says,

> Students of medicine cannot learn anything from my commen-
> taries besides what is in my medical works, where I wrote sys-
> tematically enough that the stupid man can understand. But they
> desire a knowledge of history, which makes *hoi polloi* admire prac-
> titioners in the belief that those who have read history and are full
> of bits of information understand the theoretical basis of the sci-
> ence. [CMG 5.10.2.1, p. 78]

Thereupon he quotes at length the account of the early history
of the dispute in Alexandria as it was told by Zeuxis in his commentary, "which is not now easy to find." In the process he
speaks of treatments of the text by the Empiric commentator
Heraclides of Tarentum and the Herophilean commentator
Heraclides of Erythrae. (There is no mention of Rufus.) The
information that Galen thus gives is precious for us latter-day *hoi
polloi,* and I shall quote it in the next chapter on Alexandrian
beginnings of the Corpus Hippocraticum.

Galen's next commentary, on *Epidemics* 6, returns to the manner of his commentary on *Epidemics* 2. This book, like the other,
Galen considered to be a posthumous Hippocratic work, full of
puzzles in its many brief statements and paragraphs. The puzzles, as before, send Galen to the commentators and stimulate
his talent for abuse of predecessors who tried to make sense of
nonsense. Again he apologizes for such indulgence, blaming his
readers' desire for history. Planned at the outset for eight books,
this commentary is rich in snippets of information about the

tradition, but much of it (as with the *Epidemics* 2 commentary) was lost to scholarship until recovered from Arabic translations in this century. Much of what I have said above about Galen's methods in the *Epidemics* 2 commentary applies to this also. I shall only mention here some peculiarities which are significant for our knowledge of the earlier tradition and Galen's use of it.

Galen is elusive on the subject of Hippocrates' contemporaries: when "Hippocrates" criticizes Herodicus for exercising patients who have fevers, Galen says that the question whether he means Herodicus of Selymbria or of Leontini is irrelevant. "I will discuss it thoroughly elsewhere. Now is not the time for historical research, when even leaving aside the exegesis of many passages I will be lucky to get through in eight books" (CMG 5.10.2.2, p. 171). Plato spoke of both Herodici, the author of Anonymus Londinensis of Herodicus of Cnidos. I suspect that Galen dismisses the subject here as a way of dealing with his own uncertainty. Though elusive about Herodicus, he is expansive about other subjects. He describes a hitherto unmentioned commentator in this work, Rufus of Samaria, a Jew with a large library, who did not know Greek before he came to Rome and who had the audacity to write commentary on Hippocrates. Rufus' commentary took material from everyone, said nothing original, and praised the most outlandish statements by others (CMG 5.10.2.2, pp. 212, 293). Franz Pfaff, who did the translations from Arabic and so discovered this new Rufus, made rather large claims for him as a scholar and as Galen's source that are not justified by the information Galen gives.[81]

Galen read part of Lycus' commentary on *Epidemics* 6 while doing his own and found new ways to abuse Lycus: he accuses him of stealing all his material from Marinus, differing from him only in verboseness. Galen tells us that when he found a unique reading in Lycus' commentary, he searched the shops for Marinus' books, but could not find the source (pp. 287–288).[82]

81. Franz Pfaff, "Rufus aus Samaria, Hippokrateskommentator und Quelle Galens," *Hermes* 67 (1932), 356–359. The weakness of Pfaff's thesis is pointed out by Hans Diller in *Gnomon* 22 (1950), 231–232.
82. This sounds like scurrilous exaggeration, and I suspect that Galen is automatically extending to Lycus' Hippocratic commentaries the view he had

He says also that Pelops took great pains to change obscure expressions and to explain them, but explained with such brevity that one cannot easily convict him of error (p. 291).

Incidentally in all his expansiveness, Galen gives some insights into the earlier tradition of *Epidemics* 6 and the Corpus. Zeuxis figures prominently as the oldest source of information. Galen quotes him numerous times, giving some of the flavor of his writing: Zeuxis liked to quote and criticize Glaucias; he compared Hippocrates' use of a word to Herophilus' use of it; Zeuxis, Glaucias, and Heraclides took one statement in the text as a *hermaion* (gift from the gods) for their point of view: they took it to show that Hippocrates proceeded empirically (CMG 5.10.2.2, pp. 113, 20, 174). Galen also tells us that Dioscurides placed an obelus by some words that he thought interpolated (p. 283). In a long comment on a passage describing how various fevers appear to sight and touch, Galen quotes many fifth-century poets and writers on *pemphix* (pustule) and also quotes the *Cnidian Opinions*, "which they attribute to Euryphon," for its use of the word (p. 54). Later in the same comment, he quotes two descriptions of "livid fever," one from Euryphon (work unspecified) and one from *Diseases* 2, which, he says, "seems to those around Dioscurides to have been written by Thessalus' son Hippocrates" (pp. 55-56). The quotations he gives from Euryphon and from *Diseases* 2 are virtually identical, but Galen makes no observation about it. Erroneous reading of this passage has led scholars generally to infer that Euryphon's *Cnidian Opinions* was the source of *Diseases* 2 (cf. footnote 68, above). Another odd silence occurs on the passage that reads cryptically: "What comes from the little *pinax*"; Galen says that he does not

---

given of his anatomical works. Galen attacked Lycus' reputation as an anatomist in public demonstrations and in the book *What Lycus Did Not Know About Anatomy,* where he made the point that Lycus lifted everything from Marinus, but was more long-winded and less accurate (*Scr. Min.* 2.100-101). Here he produces the new charge against Lycus' commentaries apparently out of nowhere. Further, Galen's scurrility against Lycus requires him to claim knowledge of the commentaries of Marinus, use of which he nowhere else shows evidence in extant material. I think that Galen cannot be credited here.

know what the phrase means, but that some attach it to what precedes, some to what follows, for credible reasons (p. 442). It seems very likely to me that this very passage had been the source of many of the conjectures about the manner in which Thessalus edited Hippocrates' posthumous works, which he found on *pinakes,* scraps of paper, etc. I suspect, too, that Galen knew it. His expansiveness fails to send him to the literature at this point, unfortunately, and he prefers to suppress the subject.

A work composed of snippets produces such commentary. "The urine points to the tongue." What does that mean? Perhaps it means that if you look at the tongue you can foresee what the urine will tell, but some read a different text and say absurd things (p. 286). Frivolous is perhaps not the word, but the whole commentary is at a low level of seriousness by Galen's standards. Serious-minded later Greek copyists, people who were interested in the Science Galen served, omitted much of the gossip in the text of the commentary and lost parts of it entirely, but the Arabic translation has restored much of the gossip to us.

The gossip is, of course, very precious to us, much more so in the twentieth century than lengthy, solemn medical interpretations on the ancient model would be. We now get our medicine from another tradition and read Galen for history. Because of the change in perspective, particularly our loss of partisanship in the ancient medical quarrels, our standards for historical truth are likely to differ from Galen's.

In relation to the commentary on *Nature of Man,* the next one Galen did and the last one available,[83] perspective makes me

---

83. Fragments of the commentary on *Airs Waters Places* were preserved in Hebrew, from which a Latin translation was published in the edition of René Chartier, *Operum Hippocrati Coi et Galeni Pergameni, medicorum omnium principum* (Paris, 1679), 6.187-204. German versions of the astronomical portion are given by Wolfgang Schulz, "Der Text und die unmittelbare Umgebung von Fragment 20 des Anaxagoras," *Archiv fur geschichte der Philosophie* 24 (1911), 325-334, and by H. J. Kraus, H. Schmidt, and W. Kranz, "Ein neues Hesiodfragment," *Rheinisches Museum* 95 (1952), 217-228. The published fragments belong to the first two-fifths of the work. They are not very informative about medical history. They refer to *Aphorisms* (and once to *Sevens*), but not to Galen's works. There are two references to "commen-

suspicious about what Galen is doing: I have no desire for him to succeed in establishing his point of view; rather, I desire to see through him and see what he is, in effect, if not purposely, concealing. The reason is that in this commentary Galen was defending the cornerstone of his doctrine about Hippocratic science, laid twenty-five years before. His views had been challenged, and he fought vigorously against his detractors without revealing who they were or what precisely they had said.

Galen begins his commentary by saying that long ago he wrote the work on *Elements according to Hippocrates* for a friend who was going on a trip, that he wrote no general introduction to it, and that it was widely circulated. He considered that work to be sufficient interpretation of the *Nature of Man* until his friends began to ask him to comment on all of the work, not just the statements that are crucial to the doctrine (*dogma*, the word he used for the title of the *Opinions of Hippocrates and Plato*). Now he proposes to begin his commentary with an introduction to the work on *Elements* (CMG 5.9.1, p. 3). The proposed introduction proves to be a presentation of the philosophical background for elemental theory in medicine. Galen's first question is, "What does *physis* mean?" The study of Nature is the study of the eternal and permanent which underlies phenomena. Plato said that Hippocrates' way is the proper way of studying it (p. 4). All philosophers begin their reasoning from the elements, though they have differed according to whether they defined them qualitatively or quantitatively. Having said this much, Galen refers his reader to his own work on *Medical Terminology (peri tôn iatrikôn onomatôn)* for definitions of elements and of nature.[84] Now,

---

tators" (Chartier 6.190, 200) which are not revealing. What has so far been published is largely paraphrase of *Airs Waters Places* in Galenic terms, save for an astronomical disquisition at the end. The full commentary has now been discovered in Arabic translation and will be published. See Manfred Ullmann, "Galens Kommentar zu der Schrift *De aere aquis locis,*" Corpus Hippocraticum, ed. Robert Joly (Mons, 1977), pp. 353–365.

84. The first part of the five parts of the work *On Medical Names* (or *Terminology*) has been preserved in Arabic translation and is available in German: "Galen *Ueber die medizinischen Namen,*" ed. and trans. Max Meyerhoff and Joseph Schacht, *Abhandlungen der preussischen Akademie der Wissenschaften zu Berlin,* philosophisch-historische Klasse (1931), no. 3. Apparently, we do not have

he says, the question is not terminology but those things themselves from which the first synthesis occurs. The first synthetic
bodies, Galen says, "are what Aristotle and I call *homoiomere*,"
such as bone, fat, flesh; the second synthesis of these produces
what we call the organic bodies, eye, leg, hand, and so forth (p.
6). Proper medical practice requires distinguishing diseases of
compound organs from those of *homoiomere*, as Galen does in his
works (and, as usual, he says that Erasistratus and others who do
not are only half dogmatic). Here Galen refers his readers to the
Differentiation of Diseases, "a short work," and to *Methodus
Medendi*. (I might note here that he does not know how to refer
readers to the particular area of that gross work in which he
spoke of the notion of half-dogmatism.)

But, he continues, that and other things require the line of
reasoning (*logos*) about the nature of the body that is in this work
(*Nature of Man*). Whence, he goes on (with some serious stretching of logic), one might be surprised at people who deny that the
work is by Hippocrates, "deceived as they are by the alterations
and interpolations in it" (p. 7). When he later reiterates this
point, he says that since the elemental theory of the *Nature of
Man* presents the whole basis for Hippocratic science he is surprised that people deny the work to him (p. 8).

One notes less giddy Hippocratism here than in the earlier
work on Hippocratic elements, particularly in Galen's defen-
siveness and his failure here to insist on Hippocrates' primacy.
But his general view has not changed. He goes on to say that he
will discuss the interpolations in the course of the commentary,
but he will now give quoted evidence. I will state here the evidence he quotes and the interpretation he gives it. It is either

---

the discussions of Nature and Elements to which Galen refers in the *Nature
of Man* commentary. The part that remains is a discussion of attitudes toward terms, particularly *puretos*, "fever," with considerable attention to the
literary history of the word. It was written before Galen wrote the works on
pulses, therefore before 175, but I cannot date it more precisely. In tone it is
like the rambling attacks on the Erasistrateans associated with Galen's enmity to Martialius in 163. This fact is worth noting when we are looking for
the detractors whom Galen is answering in his commentary on *Nature of
Man*.

from one of his own works or, as I suspect, from an earlier bibliographical catalogue of the Corpus Hippocraticum.

For the moment it will suffice to say what is said in the treatise (*hypomnema*) *On Genuine and Spurious Compositions of Hippocrates*, where this is said verbatim:

Section one of this book extends to 240 lines. It demonstrates that bodies of animals come from heat, cold, dry, wet, after presenting the nature of the humors. The rest is a miscellany. The first part distinguishes so-called sporadic diseases from epidemic diseases and plagues and shows the proper therapy of each type in general. The next section describes the anatomy of blood vessels, then there is miscellaneous instruction about diseases. Next there is a healthy regimen for laymen, then instruction on reducing overfat men and fleshing up thin ones, to which is added instruction about vomiting, then some discussion of regimen for children, then women, then athletes. Finally ten lines or so on diseases of the head randomly presented.

Obviously this whole book is composed from a number of works, stretched out to 600 lines or slightly less. Its first discussion (*logos*), concerning elements and humors, is entirely in accord with Hippocratic science (*techne*) as is the second which distinguishes epidemic and sporadic diseases. The anatomy of blood vessels is obviously interpolated: it is wholly bad since it does not accord with the phenomena nor agree with what is said in *Epidemics* 2. Of the rest, some is interpolated as will be specified when we explain the book to you [*soi*, the singular pronoun]. Some is worthy and well and briefly said and in accord with Hippocrates' *techne*, e.g., the discussion of the healthy regimen.[85]

85. That he is quoting from his own book *On Genuine and False Books of Hippocrates* was argued by Mewaldt, "Galenos," pp. 111–134, and has been accepted by everyone, so far as I know. Such a book was known by Hunain Ibn Ishaq, who translated it into Syrian (cf. Fuat Sezgin, *Geschichte des arabischen Schrifttums*, 3 [Leiden, 1970], 146). Nevertheless, there are grounds for doubt that he is quoting from his own book and also that the book Hunain saw was genuine). I hope to discuss it elsewhere sometime, but will say here that Galen never claims in extant works to have written such a book, and that the supposed promise to write it is in the forged commentary to *Humors*. (Mewaldt, who did not know the *Humors* commentary was forged, conjectured another promise to write the book into the *Opinions of Hippocrates and Plato*, book 5, written early in Galen's career.) If Galen is referring to someone else's book here, as I think probable, that book is likely a general listing

This is how the whole book is made up, and its first part contains, as it were, the foundation of the whole Hippocratic science. Wherefore I said I am amazed at those who estrange it from Hippocrates' ideas. Most of those who know the science of Hippocrates number it with the genuine ones in the belief that it is the composition of the great Hippocrates, some think it is by Polybus his student who took over the education of the young, who clearly departed in no respect from Hippocrates' beliefs in any of his own books. And neither did Thessalus, his son, a remarkable man also, but he did not stay in the homeland as did Polybus, but joined Archelaus, king of Macedonia. As I said, virtually all other physicians, save some few, believe that the book *Nature of Man* is by Hippocrates. And Plato himself is not unaware of it. [CMG 5.9.1, pp. 7–8]

He goes on to quote Plato's *Phaedrus* again on Hippocrates and the method of studying nature (pp. 8–9) and to say that readers of Plato, when they think it over, will realize that the method Plato describes is illustrated nowhere except in *Nature of Man* (p. 9).

But then Galen returns to discussion of his detractors and opponents, who, a moment ago, seemed not to exist, but now exist again. "When my book *Elements according to Hippocrates* fell into the hands of the general public, it was praised by all educated men, but some few of the ignorant who could refute none of the demonstrations in it, though they tried, were choked with jealousy. They thought that it was sufficient for slander to say that this is not Hippocrates' book" (p. 9). Galen goes on to say that even if it were not Hippocrates' book, the same doctrine is in the "most genuine" books, which distinguish diseases by hot, cold, wet, and dry, and from them derive their views of therapy and ages of life. He adds that since "those who slander all fine

---

of all the works of the Corpus (genuine and false) with summaries of subject matter of the sort that Galen quotes for *Nature of Man*. Such a work would be an imitation or expansion of the *Pinakes* of Callimachus and would properly accompany an edition of the Corpus. It likely comes, then, from "those about Dioscurides and Capiton." I will not carry these inferences much further without some confirmatory evidence, which I have not yet found. See also the following note.

things" have criticized the length of his writings as more than
comments, he will write a separate work proving that Hippocrates has the same doctrine in *Nature of Man* and all other writings. But he knows that the vicious slanderers will again attack
the new book: they will say that "Hippocrates does, indeed, employ elemental views of hot, cold, wet and dry, but he does not
actually have the concept." And when his book demonstrates
that all bodies are composed from those elements, they will say
that Hippocrates did not believe so, since *Nature of Man* is not by
Hippocrates. "So their viciousness never stops" (p. 10).[86]

This is as near as Galen comes to saying what discussions there
were about the genuineness of the work and its relation to Hippocrates. We do learn in the course of the commentary that
Sabinus accepted as Hippocratic everything up to the passage on
fevers (ch. 15, L 6.66), which disagrees with *Aphorisms* 4.59,
whereupon he "and most commentators" attribute all that follows to Polybus (as though, says Galen, they had read the
Aphorisms, but Polybus had not, and a student of Hippocrates
had not been trained regarding fevers [p. 87]). Of predecessors,
only Sabinus' commentary is mentioned throughout, and that
only on four passages and with considerable contempt. Sabinus
calls himself a Hippocratean, but without a dream of anatomy he
madly attempted to write commentary in praise of the chapter
on the anatomy of veins (p. 72). Galen's view is that no one could
have written the anatomy of the veins: it must have been put in
by someone who hated Hippocrates or who wanted to lengthen
the book to sell it to the Library (pp. 70–75). He gives a list of
anatomists from Diocles to Lycus, but only to assert that none of
them said eight veins came from the head (pp. 69–70). He appears to maintain a studied silence about everyone's views except
his own, save for the incidental revelation about Sabinus. I
should add, perhaps, that he tells us that Dioscurides obelized a

86. In *On His Own Books*, he says that after having written the commentary on
Nature of Man, since he had heard that some people doubted that the work
was genuine, he wrote a work in three books: *That Hippocrates Has the Same
View in Other Writings as in Nature of Man* (*Scr. Min.* 2.113). Had he written on
genuine and spurious Hippocratica in another work also, it seems extremely
likely that he would mention it.

passage (ch. 9), indicating that he thought that it was written by
Hippocrates' grandson (p. 58).

I think that it is reasonable to infer that the opponents and
detractors against whom Galen struggles in this work are the
same ones who caused him difficulties betwen the writing of the
first and second parts of *Elements according to Hippocrates.* As far
as I can determine, Galen gives us no grounds for suspecting
that he had serious literary opponents, only that people did not
find convincing his assertions about Hippocrates' innovations in
philosophy. It appears to me that such questions had never
arisen before Galen invented them. Hence his literary opponents, of whom Sabinus is the only one named, could hardly
oppose his views. But why is Galen so elusive in regard to previous commentators and their views? My view is that Galen does
have something to hide, in the sense that if he dealt with predecessors' views he would only be making the reader aware of
views which he prefers not to have to discuss and account for.

I shall take vengeance on Galen's elusiveness by noticing some
other responses to him, which have caused confusion in the
Hippocratic tradition. First I shall notice a work we will probably
never read, a Greek manuscript that came down to Hunain Ibn
Ishaq among Galen's works, which was entitled, "On the Correctness of Quintus' Criticisms of the Hippocrateans Who
Taught the Four Qualities."[87] I suspect that the book was the
product of a frustrated reader of Galen in a later period, though
other possibilities are open. However, we do have a demonstrable instance of a frustrated reader's making up the kind of
information which Galen consistently refuses to give. That is in
the Renaissance forgery, the commentary on *Humors,* which I
introduce here for two reasons: (1) the forgery's generosity
points up by contrast the genuine Galen's reticence about or
ignorance of the early tradition of the works, and (2) as long as
the commentary was considered genuine it led scholars into

87. Hunain suspected that it was not by Galen, but had not read it and so gives
no account of it in his *pinax.* Cf. Sezgin, *Geschichte des arabischen Schrifttums,*
3.137.

some false conclusions that persisted after the forgery was exposed. Here is the opening of the work:

> The ancient commentators disagreed about this writing. Zeuxis and Heraclides cast *Humors* out entirely from the genuine Hippocratic books. And Zeuxis and after him Heraclides wrote on all the books of Hippocrates. Glaucias and others say that this is by Hippocrates, but not by the great Hippocrates, author of *Aphorisms* and the other things I spoke of in *Abnormal Breathing*. But that is the view of the followers of Dioscurides and Artemidorus Capiton, both of whom had many novel views about the ancient writings. And we find various others whose views it is hard to ascertain. They say that some of this is by Hippocrates, some not. One can find many things compressed to utmost brevity, and some stretched out to more than usual length. As a result, what they perceive to have order and to be spoken in the style of the ancient himself, they so interpret as true and genuine. But what is confused or disordered or that stands out in some other way, they say is interpolated. Some were eager to find traces of extremely ancient books, written three hundred years before. I set out to study these things in the first commentators so that I could find the correct readings on the basis of the greatest number and most worthy old ones. And the results were even better than I expected. Wherefore I came to be amazed at the rashness of the commentators, who when they ought to teach what is most useful to beginning students of medicine, do everything else instead, and talk much silliness. But since some say that this writing is by Thessalus, the son of Hippocrates, or by Polybus, his son-in-law, whose writings belong to the science of Hippocrates and are not unlike either the works of Hippocrates or those which belong to Euryphon but are transmitted among those of Hippocrates, if I have more leisure some time I shall write a commentary to explain which of the books are genuine and which spurious. So I wrote in *Abnormal Breathing* about the books of the *Epidemics*, of which *One*, *Two* and *Three* seem to those who know best about them not only to be by Hippocrates, but also in their viewpoint and peculiarity to be related to one another. And the same situation was shown to apply to *Four* and *Six*, for it is agreed that Thessalus, Hippocrates' son, composed these, having found the notes of his father on skins and tablets, and that he added no

little himself. But the *Fifth* and the *Seventh* no one, I think, would consider worthy of the personal view of Hippocrates, and in my view not the *Fourth* either. But, this being so, let us proceed to the commentary on *Humors*.

The reader can see some of the sorts of misleading influence the forgery could exert, particularly its first three sentences about Alexandrian judgments on the genuineness of the work. Wellmann still used the commentary in 1891 for his still influential characterization of Alexandrian Echtheitskritik and of Galen's manner of composing commentaries.[88] Despite the removal of the evidence on which they were based, the views have persisted. It has been difficult to return to ignorance and to drop the notion of Alexandrian scholars working over the Corpus in the same manner that Galen and his immediate predecessors did, and also to drop the notion that medical men of the Alexandrian period agonized over precisely what came from Hippocrates' hand and what did not, and what precisely were his dogmas.

The forgery has had an interesting modern history also. Karl Kalbfleisch set out to edit it for the Corpus Medicorum Graecorum. In his reports to the Berlin Academy about the progress of work on the Corpus, Hermann Diels first announced that Kalbfleisch had concluded that the work was spurious, a "clumsy and careless" compilation that used printed editions of Hippocrates and Galen and Oribasius (SBA 1913, p. 116).[89] Two years later he reported that Kalbfleisch had concluded that a "Byzantiner" had composed the work and that among the pieces of the mosaic stuck together with words from the compiler were some fragments from lost works of Galen

---

88. Wellmann wrote on medicine for Franz Susemihl, *Geschichte der griechischen Literatur in der Alexandrinerzeit,* I (Leipzig, 1891), 777–828. He was, of course: unaware of the material that has since been recovered from the Arabic, including the genuine commentary on *Epidemics* 2, which might have led him to the correct conclusion.

89. Hermann Diels, *Sitzungsberichte der Akademie der Wissenschaften zu Berlin* (1913), p. 116.

(1915, pp. 92–93). But then he discovered the source of those; Diels reported that the forger had used a Latin translation of Moses Maimonides' *Aphorisms*, which appeared in 1489. This confirmed the Renaissance date of the forgery and accounted for the pieces from Galen's commentary on *Humors* that the forger had used: they are in Maimonides. Kalbfleisch, according to Diels, was finishing work on the text, which was to be published as a demonstration of the forger's methods (1916, p. 138). In 1932, Hans Diller spoke of Kalbfleisch's edition as forthcoming (*Warderarzt und Aitiologe*, p. 151). In his 1965 bibliography, Konrad Schubring reported that the material for Kalbfleisch's edition was destroyed (K 20.XLIX).

I can summarize under a few headings and in simple terms what I have shown about Galen's relation to the Hippocratic tradition and the Corpus Hippocraticum.

1. Galen's immediate medical circle and teachers all used and responded to the Hippocratic Corpus in their medical instruction, but with quite varied attitudes. Sabinus and the others who called themselves Hippocrateans seemed scientifically backward to Galen.

2. Galen's version of Hippocratic science and its tradition is in large part his own, a projection of his concerns onto history. While his medical system was put together out of Hellenistic medical developments, his peculiar Hippocratism was fashioned largely as a rhetorical and ideological patina for it. His claims about Hippocrates' original philosophical and scientific system were put forth for the circle of intellectuals in Rome, phrased in terms relevant to them.

3. Galen's treatment of medical men from Herophilus to his own time as enemies or followers of Hippocrates can generally be ignored save where (as with Asclepiades and the Empiric commentators) there is supporting evidence.

4. Echtheitskritik on the works of the Hippocratic Corpus had preceded Galen, but not by much. Commentary and glossography from the Hellenistic period were available to him in excerpts or in the original works. The worth and fullness of

Galen's testimony about this material varies according to the material he was using, but by putting together the bits of evidence he offers we can get an idea of the kind of work that had been done on the Corpus in various earlier periods. In what follows, I shall add to Galen's testimony what independent evidence we have and will attempt a chronological reconstruction.

3

FROM HIPPOCRATES
TO GALEN

We have seen what Hippocrates probably wrote and what
Galen did with the Hippocratic tradition that he inherited. What
remains is to make the connection between the two: how did the
tradition that Galen inherited come into existence? Specifically,
how was the Corpus Hippocraticum assembled and how was it
read and by what stages did the "father of medicine" receive his
attributes? Having dealt with the nature of Galen's evidence, we
are, I think, in a position to revaluate the subject. Much of what
we deal with is highly speculative because the information that
comes to us is so sparse. What I shall do is pose the questions (1)
How did people describe their own notions of medical practice,
and (2) How did they relate themselves to the tradition, particularly to Hippocrates? Where there is no evidence, we can apply
skepticism to conjectural reconstructions that have been based
on patterns of thought such as were traced in Chapter 1 and
inevitably speculate in return.

For the sake of clarity I shall offer here a preliminary sketch
of the important trends that I shall find and the conclusions I
shall reach.

1. In the period of creative, "dogmatic" medicine in the
fourth and third centuries B.C., physicians of whom we have
information attempted to establish a sound logical foundation
for a coherent medical system. Individuals differed widely in the
proposals they offered, but they seem to have shared the wish to
arrive at sensible statements about causes of the phenomena of
health and disease. They preferred reason to authority or tradi-

tion. There is no evidence, I shall say, that any of them acknowledged Hippocrates or any other predecessor as an authority to be followed or to be dethroned. A "school" or schools of logical medicine came into being in Alexandria in the wake of the most influential of the dogmatics, Herophilus and Erasistratus.

2. In the period roughly from 225 to 50 B.C., while the results of the work of logical medicine persisted largely as received dogma, there was also a movement of reaction and reform by Empirics. They accused their predecessors of practicing medicine in the abstract, away from the bedside and the facts of disease and health. They accompanied their insistence on experience rather than theory with a revival of those archaic, predogmatic medical works that had been collected by the library at Alexandria and attributed en bloc to Hippocrates. They admired the "Hippocratic" works as records of useful medical experience, unaffected by dogmatic theorizing. Empirics thus wrote the first medical commentaries on works of the Corpus Hippocraticum, making use of the dictionaries of archaic terms which had been written earlier.

3. In the period of the first centuries B.C. and A.D. other important reactions to preceding medical theory and practice occurred. Two new varieties of dogmatic, or logical, medicine became prominent. Asclepiades, inspired by atomistic, probably Epicurean views, presented a new logical basis for judging medicine. He attempted to revise and correct preceding theory and practice, that of Hippocrates, the Dogmatics, and the Empirics. The Methodists, whose theories later developed out of Asclepiades' work, followed his lead in considering Hippocrates venerable but in need of radical correction according to proper modern theory. The other dogmatic school, the Pneumatics, appear to have developed their logical basis for medicine out of Stoic philosophy. Their attention was given to revising dogmatic medicine, and they did not concern themselves with Hippocrates, either to correct him or to attribute dogma to him. Nevertheless, the Hippocratic works were widely read. In popular romance, and even in sophisticated accounts, Hippocrates became *the* father of medicine, who was to be revered, though it was not necessary to know in detail what his doctrines were.

4. Toward the beginning of the second century A.D., Dioscurides and Capiton collected the Corpus Hippocraticum and presented it in its first literary edition. They also decked it out with theories as to the provenance of its very various works. Shortly later Rufus of Ephesus wrote his commentaries which excerpted the works of the earliest Empiric commentators on the Corpus. Subsequent medical work and Hippocratic interpretations by Galen's teachers and teachers' teachers included, as we have seen, on the one hand (Marinus, Quintus, *et al.*) revival of anatomical research, and, on the other (Sabinus *et al.*) a new dogmatic Hippocratic medicine. Sabinus' commentaries integrated the historical conclusions of Dioscurides and Capiton with a reading of Hippocrates as a teleological theorist in humoral pathology. Marinus, perhaps followed by Quintus, attributed anatomical investigation at least, perhaps also empirical heuristic method to Hippocrates. As we have already seen, Galen integrated those two points of view and advanced them both, and also probably a Stoic, pneumatic version of Hippocrates that his teacher Aephicianus had developed. Thence, Galen attributed his own syntheses to Hippocrates and adjusted the other facts of medical history to conform to his outlook. There was indeed, change and development in the six centuries before Galen, in medicine and in ideas about Hippocrates.

FROM HIPPOCRATES TO ALEXANDRIA

We begin with Hippocrates' own time. Since we need not assume, as Galen did, that all medical work and writing was either drawn from or opposed to Hippocrates, we can ignore Ctesias' statement (if he made it) that the dislocated hip cannot be successfully reduced. An itinerant physician and historian in the Near East. Ctesias is known to have spoken of progress in medical practice in at least one respect: he said that proper dosage of the dangerous hellebore was better understood in his own time than in that of his father and grandfather (Oribasius 8.8).[1]

1. Oribasius, *Collectionum medicarum reliquiae,* ed. Johannes Raeder (Leipzig, 1928–1933).

Had Ctesias left any serious account of medical practice or medical literature, it should have survived. Probably his appearance
in medical history is entirely the product of later desire for controversy.

Dexippus and Apollonius,"students of Hippocrates," are, if we
disregard Ctesias, our closest contemporary contacts with Hippocrates. They are said to have used a severely restricted regimen for fever patients. But Galen, who gives this information,
infers, probably rightly, that Erasistratus' report of that fact is
based on hearsay, not on their writings. Only Dexippus is known
to have left writings: one book on medicine and two books on
prognosis,[2] and the titles are all we know of them. Dexippus was
not influential in medicine. But, like Hippocrates, he later became a figure in the mythical political history of Cos: he, so the
story went, saved the island from fear of war by healing the sons
of the king of Caria.[3] More substantial is the account of his
"doctrine" in the Menon papyrus (*Anon. Lond.* 12.8).[4] His doctrine resembles those of the other dieticians, and, typically,
Menon attributes a peculiarity to him: he thought that disease
was the result of residues of nutriment from overeating, as well
as excess of heat, cold, or the like. But he "seems to differ from
the others I mentioned," says the papyrus, "because he says bile
and phlegm melt and produce sera and sweats, which thicken to
pus, mucus, and rheum." Erasistratus inferred more of Dexippus and Apollonius: they gave so little drink to fever patients
because they thought liquid was fuel for the fever.[5] Though one
might be tempted to say, "Ah yes, just like the view of Hippocrates in his single work *Regimen*," it would be as relevant to cite
the old drinking song, "the sun drinks the sea," as a parallel for
their view, if they held the view. The view was extremely common, and grand theories have been constructed on hasty as-

2. This information comes from the *Suda,* a tenth-century dictionary. Its source
   is not clear. Cf. Max Wellmann in *RE* 5.294–295, s.v. Dexippos.
3. This information is also from the *Suda.*
4. *The Medical Writings of Anonymus Londinensis,* trans. William H. S. Jones (Cambridge, 1947).
5. CMG *Suppl. Or.* 2, p. 107.

sumptions about influences.[6] Hence, we know little of Dexippus and Apollonius except that they lived, were known as Hippocrates' students, and became bywords for restricted regimen. That Coan myth made a political hero of Dexippus is itself of interest, since similar myths about Hippocrates were important to the later Hippocratic tradition. For the moment we shall ignore the question of Polybus, or Polybius, the supposed son-in-law of Hippocrates, and also Hippocrates' sons and grandsons. We shall have laid the groundwork for a fairly extensive skepticism in relation to them (below, pp. 218-220).

The next important figure is Diocles, who worked in Athens and wrote in the Attic dialect, a factor which may have made him more interesting to later medical men than he would otherwise have been. "Second in time and fame to Hippocrates," "*Sectator Hippocratis*," are among the descriptions that come to us. He is important in my investigation because he is fairly well represented in fragmentary remains: he was famous and influential enough to be quoted often by Galen, Oribasius, and others. Diocles provides the best source of information (parallel to the philosophical sources, Plato, Aristotle, and the fragments of Menon) for thought about medicine in late classical Athens.[7]

In subsequent tradition, Diocles stands at the head, or second to Hippocrates, in lists of "logical" or "dogmatic" physicians. Not untypical is the list of dogmatics in the pseudo-Galenic *Introduc-*

6. Wellmann, *RE* 5.294-295, tries to relate Dexippus to "schools" with odd results. Karl Deichgräber, *Die Epidemien und das Corpus Hippocraticum* (Berlin, 1933), p. 168, can relate him to a Coan school, of course. Galen is furious about the subject because, if what Erasistratus said of Dexippus and Apollonius is true, they disagreed with that eminently "Coan" work *Regimen in Acute Diseases* (Galen in CMG 5.9.1, pp. 256, 277). Cf. Hermann Grensemann, *Knidische Medizin I* (Berlin, 1975), 205-217.
7. The fragments were collected by Max Wellmann in *Die Fragmente der sikelischen Aerzte* (Berlin, 1901). Werner Jaeger, in *Diokles von Karystos* (Berlin, 1938) and in various articles, tried to bring Diocles into connection with Aristotle and to date Diocles' work in the third century B.C., a century later than Wellmann had dated him. Fridolf Kudlien, "Probleme um Diokles von Karystos," *Sudhoffs Archiv* 47 (1963), 456-464, gives an account of the reactions to Jaeger's work and of the status of the question.

tio, sive Medicus (K 14.683): Hippocrates, Diocles, Praxagoras, Herophilus, Erasistratus, Mnesitheus of Athens, Asclepiades of Bithynia. That list was made after 50 B.C., since it contains Asclepiades. In questioning when such lists began to be composed and what the first ones look like, I can infer only that the Empirics first declared themselves a *hairesis,* a sect or school in medicine. The term *logikos* may have been used earlier by physicians to describe their orientation (though I have no proof), and the detractors named the school *dogmatikoi* as a term of abuse ("true believers," perhaps). "Logical" and "dogmatic" later are synonyms opposed to "empiric," and still later to "methodic." In Diocles' time, therefore, I assume that there were no Dogmatics or Empirics, just as there were no Christians or Epicureans, but that later interpreters could find tendencies in Diocles which were clearly of the type. I conclude that Hippocratism, too, is a later phenomenon, first sponsored by the Empirics, who claimed that Hippocrates was empiric. Later still, the list of Dogmatics, with Hippocrates at its head, evolved from attempts to write the whole history of medicine in the light of Empiric categories. This is shown, for example, in Celsus' list of those who pursued medicine: after the early philosophers came Hippocrates, Diocles, Praxagoras, Chrysippus, Herophilus, and Erasistratus; later came the Empirics, later still came Asclepiades, and then the Methodists, who developed his point of view (Celsus, *Proemium* 6-11).[8] Just as there is a progression by genre in Greek literature, from epic to lyric to tragedy, as it was viewed in retrospect, so there was a progression in medical work, as it was viewed in retrospect, when attempts at medical history began. I think that Diocles' relationship to Hippocrates is an imaginative product of these later lists, then, which are an attempt to create a history of medicine.

There was a sort of history of medicine in or near Diocles' time: Menon started it by asking what causes people adduced to explain diseases. Menon had not conceived of the categories dogmatic versus empiric. His categories separate those who

8. Celsus, *De Medicina*, ed. and trans. W. G. Spencer, Loeb Classical Library (London, 1935-1938).

picked different kinds of causes for diseases, elemental versus excremental. He shared Diocles' orientation to causation as the central subject in medicine, but he had no inkling of the schools that later developed. Wherever in time we place Diocles—early, mid, or late fourth century, that is, before or after Menon— these conclusions are appropriate. After the Empirics had done their work, dogmatic methodology could be praised in opposition to empiricism, as it was by Hegetor.[9] But the early "dogmatics," apparently, were innocent of controversy about dogmatism, while interested in the questions that defined the dogmatic school. So much is clear, but particular points need further argument because historians of medicine have taken some of Diocles' cautionary words about the limitations of the *logos* and turned Diocles into a fledgling empirical scientist. I know of no English translation of Diocles' methodological fragment, so I offer one here in order to discuss the context in Galen where it appears and to contrast the quote with what Galen says about it.

Galen opens his work on *Faculties of Foods* with a long, rhetorical discussion of method in the subject, the purpose of which is to present himself as uniquely analytical and logical in the subject, but also (and somewhat indirectly) to say that he shares the subject with Hippocrates. Diocles is introduced as a foil:

> But Diocles, though he was a dogmatic, wrote the following verbatim in Book 1 of the *Hygiene to Pleistarchus:*
> "Those who suppose that foods which have like humors or smells or heat or anything else of the kind all have the same faculties (*dynameis*) are mistaken. One could point out many dissimilar results which come from such similar things. One must not suppose that everything which is laxative or diuretic, or which has some other *dynamis* does so because it is hot, cold or salty since the sweet and bitter and salty, et cetera, do not in all cases [prove] to have the same *dynameis*. Rather, one must consider that the whole nature (*physis*) is responsible (*aitios*) [for what usually occurs for each].[10] So considering, one is least likely to miss the truth. Those who think they must give a cause in each case why a thing is

9. As related by Apollonius Citensis, CMG 11.1.1, pp. 78–80, cf. below, p. 213.
10. The text is shaky here, but this seems the likely meaning.

nourishing or laxative or diuretic or something of the kind seem
to be ignorant, first that such things are frequently not essential
for use (*chreai,* plural), second that many things are in some fashion like first principles (*archai*) in nature, so that they do not
admit of reasoning about causes (*hyper aitiou logos*). And also, they
sometimes err when, seizing on the unknown, the unacknowledged, and the implausible, they think that they give an adequate
account of the cause. We must ignore those who etiologize so, and
those who think that a cause must be offered for everything. We
must rather trust those who reached understanding from experience (*empeiria*) through much time. But we must seek a cause for
what we accept when it will make what we say more intelligible or
credible."

This is what Diocles says, because he is of the opinion that one
learns the *dynameis* of nutriments from experience (*peira*) only
and not from indication (or demonstration, *endeixeis*) from temperament (*krasis*) or from humors. And though there is indication
of another sort, according to parts of plants, he did not mention
that.   [Galen, CMG 5.4.2, pp. 201–203]

Galen quotes Diocles to show that Diocles did not succeed in
creating a science in Galen's sense. In his enthusiasm he overinter-
prets the words he quotes: "only from experience" is not Diocles'
point, rather lack of gullibility is: one should trust descriptions
of effects based on long experience in preference to superficial claims to analytical insight. In his own long introduction,
Galen agrees that one must start from experience, but to be
scientific one must end with a description of *kraseis* of foods
which can be related systematically to the *krasis* of the patient's
body and to the illness. Having proved that Diocles did not man-
age such a science, Galen slyly suggests that Hippocrates was
more dogmatic than Diocles by giving a sprinkling of Hippocratic quotes in the argument (such as, "Exercise, food, drink, sleep,
sex, all in moderation," and "To achieve equality in food and
drink requires experience," from *Epidemics* 6.6.2, L 5.304, and
Epidemics 2.2.11, L 5.88). At the climax of his introduction,
Galen works up to invocation of *Regimen* as his predecessor in
the science:

Whence it is extremely necessary to analyze the *kraseis* of men and of foods for the present inquiry. Men's *kraseis*, how many these are and how one should diagnose them is discussed in my book *Temperaments*, and in those on the Faculty of Drugs I discussed the *kraseis* of drugs. Now in the present work it will be appropriate to speak of the *kraseis* of foods, as in the book called *Regimen*, which was written by Hippocrates according to some. . . . [Here follows the passage on authorship and transmission of *Regimen* discussed above on page 59, after which Galen resumes.] Let all this be said by the way. Whichever of the men we have named wrote it, it appears to refer regimen in diet to a general method. [CMG 5.4.2, pp. 212–213]

But Galen does not acknowledge that *Regimen* expressed much the same view that Diocles does:

Those who have undertaken to treat in general either of sweet or fat or salt things, or about the power of any other such thing, are mistaken. The same *dynamis* does not belong to all sweet things nor to all particulars of any other class. For many sweet things are laxative, many binding, many drying, many moistening. . . . Since, therefore, it is impossible to set forth these things in general, I will show what power each one has in particular. [2.39, L 6.534, Jones's translation]

Diocles is more abstract in his statement: not all salty, bitter, and other things have the same *dynamis*, he says; rather, the entire nature of the food accounts for its action, that is, is the cause. Both *Regimen* and Diocles reject explanation by superficial category. Scholars, however, have taken Galen at his word. Carl Fredrich's influential *hippokratische Untersuchungen* considered Diocles an Empiric who criticized *Regimen*, but who himself followed Hippocrates' doctrines in *Ancient Medicine*.[11] Wellmann, Deichgräber, and lately Fridolf Kudlien have followed in various degrees.[12] Werner Jaeger related Diocles' em-

11. Philologische Untersuchungen 15 (Berlin, 1899), 171–172, 217 ff.
12. Wellmann, *Die Fragmente der sikelischen Aerzte*, p. 163, citing Fredrich; Karl Deichgräber, *Die griechische Empirikerschule* (Berlin, 1930), pp. 274–275:

piricism to Aristotles' methodology in *Nichomachean Ethics*
(1094b 11, 1095b 1–8, 1098a 27–32), specifically to the notions
that different sciences aim at precision (*akribes*) in different degrees and that in such matters as the Good, "what is" is as far as
one need inquire, "why" being unnecessary.[13] There does seem
to be general similarity in topic, although Aristotle and Diocles
are not so close as to make direct relationship demonstrable. In
any case, Diocles' talk about method in fact demonstrates his
attention to "causes" as a center of concern in medicine and
indicates that he does not wish to be fatuous or gullible in talking
of causes. We have fragments of a work by him entitled *Pathos,
Aitia, Therapeia* (*Frag.* 37 ff. Wellmann), which title was translated by Caelius Aurelianus as *de passionibus et earum causis et
curationibus* (*Frag.* 29). The longest fragment (*Frag.* 43), a quotation with comments by Galen from *Affected Places* (K 8.185ff.),
concerns melancholy. The quotation and Galen's remarks indicate that Diocles' work presented the symptoms of the disease,
then specified the causes of the symptoms. The rationale of the
cure depended on perception of the cause, which seems to define dogmatic or logical medicine properly so called. That Diocles stands at or near the head of it seems very likely from the
later tradition and from the little available evidence. What then
of his "Hippocratism" or lack of it?

"vertritt er deutlich den Standpunkt des aristotelischen Empirismus" (he
clearly represents the point of view of Aristotelian empiricism). Kudlien,
"Probleme um Diokles," p. 461: "Wir besitzen von diesem ein für seinen
methodologischen Standpunkt hochst aufsclhussreiches wörtliches Fragment... in welchem sich Diokles deutlich als Empiriker erweist, der aus-
schliesslich die individuellen Gegebenheiten zu berücksichtigen verlangt
und jede spekulative Aetiologie ablehnt" (We have from Diocles a direct
quote which is very informative for his methodological standpoint, ... in
which he shows himself an empiric who demands that one consider exclu-
sively the individual phenomena and who rejects all speculative aetiology).
After overinterpreting Diocles' empiricism in this way, Kudlien expresses
doubt that such a person could hold a schematic four-humor theory and a
speculative pathology as the doxographies say. Doubt about the doxographies seems proper on various grounds, but it is wrong to exaggerate
Diocles' "empiricism" and to insist that he would infer what we infer from it.
13. Jaeger, *Diokles,* pp. 40–51.

Hippocrates may well have been for Diocles what he was for Plato, a known figure somehow distinct from other dietitians and teachers. But I have argued above that Plato found Hippocrates interesting precisely because of what he quotes from him, words Plato quotes because they are reminiscent of his own philosophical posture. Diocles shows no such interest in the preserved fragments. I think I may even indulge in the *argumentum ex silentio:* If Diocles had discussed Hippocrates, for good or ill, Galen probably would not have let it pass unnoticed—one can well imagine Galen's fervent compliments to Diocles' good taste or, alternatively, indignant condemnation of his vicious jealousy. I am confident that Diocles was silent about Hippocrates, although he may have read *Regimen,* to which his dietetics and his methodological statement show similarities. But there was already a dietetic tradition, which *Regimen* sets out to transmit: "I have resolved to accept what my predecessors have well thought out," although "none of my predecessors has successfully treated the whole subject" (L 6.466), and to add to it: "I shall explain also the nature of those things which none of my predecessors has even attempted to set forth" (L 6.48). There is no evidence that any of the peculiar material in *Regimen,* such as *prodiagnosis,* was noticed by Diocles, and therefore no grounds for a Hippocratic tradition in relation to Diocles and early dogmatism. I might add, though by now I hope it is unnecessary, that he shows no evidence of awareness of competing "schools." If the reader is convinced, he will look with sympathy on this next paragraph in which I try to explain away the few contrary indications.

Diocles is reported to have made pronouncements on three controversial subjects that could relate him to Hippocrates: setting the hip joint, prognosis from worms, and the five, seven, and nine day periods of fevers. Of the last he said, "You cannot say on grounds of what signs or humors there is a fifth, or seventh, or ninth day period," a statement which Galen for rhetorical purposes takes to be criticism of Hippocrates (CMG 5.10.1, pp. 112–113). Diocles' view about prognosis from worms is juxtaposed by Soranus to an opinion which he says is that of Hippocrates, but does not occur anywhere in the Corpus Hippocraticum. It is not clear how the error of attribution occurred, but other considerations suggest that Soranus was drawing on a doxography arranged dialectically—a says b, c says d, but e says f, a sequence to be followed by the writer's triumphant refutation of them all (see the discussion in connection with Asclepiades, below, p. 224). Such errors can creep in if the source of the doxography is rhetorically different from the later version and needs translation: "Some say, others say, but I say." There are many examples of such rhetorical flights among ancient medical writers, most of them early. Diocles is exemplary of the style everywhere. He never names a predecessor to refute, but speaks always of "those who . . ." as he does in the methodological fragment. Hence, some process of adding names to the nameless "some" and "others" could have produced error in the doxographies. I can go no further, but I oppose that much to the conjectures that parts of the Hippocratic Corpus which Diocles criticized have been lost, an idea once popular.

In the lists of Dogmatic physicians, Praxagoras was the next medical writer whose influence persisted. In spite of the existence of Fritz Steckerl's useful collection *The Fragments of Praxagoras of Cos and His School* (Leiden, 1958), the preserved material gives little insight into his verbal style and none into his method of expressing his concerns or theories or of the way he treated predecessors. Repeated use of his name by Galen and others in lists of dogmatics tells little. The information on his therapeutic procedures and definitions of diseases, found largely in medical doxographies, is too inconclusive even for the conclusions Steckerl draws, I believe. Praxagoras' classification of ten humors was influential, especially the name *hyaloeides* for glassy phlegm (*Frag.* 22–25). Galen tried to reconcile the four-humor theory which he attributed to Hippocrates with the ten humors of Praxagoras (*Frag.* 21). Praxagoras offered some kind of *logos* about the flow of pneuma in the healthy body and its blockage in diseased conditions, he thought that pneuma was transmitted by the arteries, which become nerves in the extremities, and he offered some kind of *logos* for the pulse (*Frag.* 26–31). Galen calls him a shameless sophist and enemy of Hippocrates because of his ignorance about the nerves (*Frag.* 11), but that is one of Galen's bad moods. Galen is equally certain that Praxagoras' explanation of cotyledons in the womb represents the view of Hippocrates (K 2.960), but he made that statement before he developed his Hippocratism and attacked others in its name. In sum, then, all the available information about the relation of Hippocrates and Praxagoras comes from Galen's rhetorical formulations and is therefore worthless. Of the man himself we can get only the dimmest notion. Most notably lacking is any description by the man himself of his reasoning and his notion of medicine. Besides therapeutic works, he wrote an *Anatomy* (*Frag.* 10) and perhaps a *Physics* from which comes his statement that cotyledons are the mouths of veins and arteries in the womb (*Frag.* 13). Such titles seem proper in the wake of Diocles and Aristotle and may hint at what medical questions Praxagoras posed. But he did not pose them so interestingly that anyone in later time quoted him or discussed his orientation to the past in medicine. Steckerl has overenthusiastically proposed finding close correspondences between *Ancient Medicine,* Menon's report of Hippocrates' doctrine of *physai,* and the bubbles (*pompholyges*) on the basis of which Praxagoras apparently explained some pathological states.[14] I cannot find grounds for such interpretations. We must, since we have no grounds for denying it, consider it possible that Praxagoras of Cos had knowledge of views about Hippocrates of Cos. Steckerl has good reason for seeking such evidence, but Praxagoras is too shadowy. In addition, the case for my argument from silence is strong: Galen knows of no direct link between the two men, no mention of the one by the other.

More information exists on Herophilus and Erasistratus, of the next generation.[15] They are the most influential physicians

14. Fritz Steckerl, *The Fragments of Praxagoras of Cos and His School* (Leiden, 1958), pp. 38–44. Steckerl finds Praxagoras "a rather earthy personality, a hard boiled materialist" (p. 33). I cannot claim such insights. The best recent (1954) work on Praxagoras is Bardong's article in *RE, s.v.* Praxagoras (1).

15. For general descriptions of the two men I refer the reader to the usual handbooks and dictionaries. John F. Dobson attempted to describe their "systems" and offered translations of the then known fragments in "Herophilus of Alexandria," *Proceedings of the Royal Society of Medicine,* Section of the History of Medicine 18 (1925), 19–32; and in "Erasistratus," ibid.

of antiquity before Galen, with the possible exception of
Asclepiades (and, of course, Hippocrates as represented by the
tradition we are pursuing). It appears that they were influential
because their work set a standard for medical thought in subsequent centuries. Yet, despite all we know of them, they remain
mysterious in important ways. Peter M. Fraser has recently
found it possible to argue again that neither Erasistratus nor
Erasistrateans worked in Alexandria and to deny that Erasistratus pursued anatomical research. Fridolf Kudlien found it
possible to write an article, "Herophilus und der Beginn der
medizinischen Skepsis," in which he attempted to put
Herophilus in the tradition of philosophical skepticism that produced Empiricism.[16] The perennial argument about whether
one or both practiced vivisection on humans is less central to
understanding of their medical thought, but symptomatic of our
ignorance.[17] We can list many things first seen and described by
each, particularly by Herophilus, who contributed several new
anatomical terms. There is greater difficulty in trying to extract
from their statements a sense of what they thought medicine was
and how they oriented themselves to the past. That is my pur-

---

20 (1927), 21–28. Robert Fuchs summarized his own work on Erasistratus in
"De Erasistrato capita selecta," *Hermes* 29 (1894), 171–203. Peter M. Fraser,
Ptolemaic Alexandria (Oxford, 1972), offers a good account of medicine in
Alexandria, with bibliography and many quotations from original sources.
Because he argues that Erasistratus did not work at Alexandria, he ignores
him in *Ptolemaic Alexandria,* but makes up for that in part in "The Career of
Erasistratus of Ceos," Istituto Lombardo di Scienze e Lettere di Milano,
Classe di Lettere e Scienze Morale e Storiche, *Rendiconti* 103 (1969), 518–
537. Geoffrey E. R. Lloyd, "A Note on Erasistratus of Ceos," *Journal of
Hellenic Studies* 95 (1975), 172–175, has argued again for his association with
Alexandria.

16. *Gesnerus* 21 (1964), 1–13. Kudlien appears to be playing with terminology to
a great extent. Unanimous designation of Herophilus as dogmatic in antiquity makes it appropriate to pose the question of what his dogmatism
consisted.

17. John Scarborough, "Celsus on Human Vivisection at Ptolemaic Alexandria,"
Clio Medica, 11 (1976), 25–38, has reviewed the subject and concluded that
stories of vivisection were probably false. I am unwilling to conclude on the
basis of the general likelihoods adduced by Scarborough that such experimentation never took place and that the discussions Celsus reports have
no basis.

pose here. I shall content myself with dealing with problems that are directly relevant to my subject.

Herophilus apparently discussed Praxagoras by name in order to refute him at the opening of his book about pulses. Galen, who gives that information, calls Praxagoras Herophilus' teacher in the same passage (K 8.723). Such specific discussion of predecessors appears still to have been rare in Herophilus' time, but became increasingly popular afterward. I mention, only to dismiss them, modern scholarly myths that Herophilus collected the Hippocratic works, commented on them, and rejected them in favor of empirical research; practically the only thing that has not been conjectured is that he wrote some of the Corpus. Admittedly there were some few things which appeared to give evidence of some attention by Herophilus to Hippocrates: the erroneous text of Galen that listed "Herophilus and Bacchius" rather than "the Herophilean Bacchius" as explicator of the *Aphorisms* (K 18 A.187)[18] and a common misreading of another Galenic text which quotes Zeuxis' comparison of Hippocrates' use of a word (*nêpia*) with Herophilus' use of it. By misplacing the quotes, editors read into the text a statement by Herophilus in which Hippocrates was mentioned.[19] That Herophilus was a student of Praxagoras of Cos provided a link between Cos and Alexandria, so the frailty of the evidence was not noticed for a considerable time.

Only one apparent indication of a relationship between Herophilus and Hippocrates remains. In part it confirms my principle that we can infer from Galen's silence people's failure to mention Hippocrates or the Corpus. Galen can attribute to Herophilus only a single, very indirect criticism of Hippocrates: at the opening of his commentary on *Prognostic*, Galen says in an aside that the statements of Herophilus' followers, who distinguish *prognosis* from *prorrhesis*, are not only useless and im-

18. There is a similar error in the introduction to Galen's *Glossary*, which in Kühn's text says that Herophilus and Bacchius explained only the rare words in Hippocrates (K 19.65).
19. See CMG 5.10.2.2, p. 20. Hans Diller, in his review in *Gnomon* 22 (1950), 231, points out the error and describes its persistence.

proper but sophistical and false (CMG 5.9.2, p. 203). Shortly
later he brings the subject up again to say, "Perhaps I should not
have mentioned Herophilus earlier. It is better to expose the
truth as quickly as possible for those who are interested in the
Science and not to exercise them doubly—once teaching the
nonsense of the sophists and again refuting it. I think that it
would be better if I proceed with what is useful in this commentary and at greater leisure consider Herophilus' criticisms of
Hippocrates' *Prognostic*" (p. 205). Again in the commentary on
Prorrhetic, written some years later, Galen observes, aside, that
we cannot accept Herophilus' legislation that *prognosis* differs
from *prorrhesis* by more or less certainty (CMG 5.9.2, p. 1), but in
this case he makes no gesture toward responding to Herophilus.
These two commentaries indicate that Galen believed that
Herophilus (cited, probably, by Herophileans) wrote against
Hippocratic prognosis, but if his words are regarded with a
skeptical eye, Galen seems to know only about the distinction
between *prognosis* and *prorrhesis*. Scholars who were sure that
Herophilus was a Hippocratic commentator found it sufficient
to attribute that distinction to his commentary on *Prognostic*.[20]
Without further information, we might properly say that Galen
had blown up a perfectly normal verbal distinction made in a
Herophilean work, or even by Herophileans, into an imaginary
historical quarrel. But Caelius Aurelianus' translation of
Soranus gives further information. *Chronic Diseases* 4.8[21] says
that ancient physicians announce the course of diseases on the
basis of worms, but they differ. Hippocrates in *Prognostic* says
that expulsion of dead worms is always a fatal sign. But Diocles
holds that vomiting worms is of no significance and that expression of them from the anus is a bad sign only if the worms are
healthy and full of blood. Herophilus, in his book against the
prognostic, holds that expulsion of worms whether dead or alive

20. See, for example, Gossen *RE* 8.1 (1912) 1105–1110, and Dobson,
    "Herophilus," p. 19. While scholars imagined that Herophilus was the pious
    collector of Hippocratic works, however, they could not easily explain his
    apparent rejection of Hippocratic doctrine.
21. Caelius Aurelianus, *On Acute Diseases and On Chronic Diseases,* Ed. and trans.
    Israel E. Drabkin (Chicago, 1950), p. 88.

is not a bad sign. Interestingly enough, the views attributed to Hippocrates here by Soranus appear nowhere in the Corpus Hippocraticum. Later in the same passage Caelius (that is, Soranus) cites Chrysippus, a follower of Asclepiades, who said that Diocles and Hippocrates do not disagree and who cited *Prognostic* 11 ("it is favorable when round worms are passed near the crisis") as evidence. Two things appear to have occurred in the transmission of this doxographic list: first, by a normal error of interpretation, a view about worms which Diocles rejected was attributed to Hippocrates and then specifically to the *Prognostic*. Chrysippus made his correction before he passed the list on. Second, a name was made up for the book in which Herophilus wrote of worms. We do not know where he discussed *prognosis* and *prorrhesis,* though the subject would seem appropriate to his book *Against Common Opinions (pros tas koinas doxas).* That he wrote a book in opposition to Hippocrates' *Prognostic* seems unlikely in the extreme.

Galen enjoys twitting Herophilus for half-dogmatism (for example, K 9.728, K 10.184, CMG 5.9.1, p. 507). He says of Herophilus' description of pulse and rhythm for diagnostic purposes, "He wrote more in the manner of a person who is giving an account of observation and experience rather than one teaching a logical method."[22] In the work *On Precipitating Causes,* Galen uses Erasistratus for his rhetorical opponent primarily, but reserves a swipe for Herophilus at the end. (He mentions Herophilus earlier as one who admits causes *ex hypothesi*: "Some say there are no causes; some, like the Empirics, say that there is doubt about their existence; some accept them hypothetically, as Herophilus; and others, followers of Herophilus, reject precipitating causes as unintelligible.")[23] Galen seldom lets anyone escape with a small bruise, and he returns to Herophilus later. Herophilus, Galen says, argued that causes could not be demon-

22. K 9.278. C. R. S. Harris, *The Heart and Vascular System in Ancient Greek Medicine* (Oxford, 1973), pp. 192–193, gives a clear account of Galen's argument in this regard.
23. *De causis procatarcticis,* ed. and trans. Kurt Bardong (Leipzig, 1937), CMG Suppl. 2, pp. 41–42.

strated, but he used them (presumably in his medical reasoning) because, he said, men believe in them (Galen calls that attitude cowardly). Finally, Galen quotes him as saying, "Either corporeal is the cause of a body or the incorporeal of the incorporeal, and the other possibilities to be found by the division. But none of them proves to be so. Whence there is no cause whatsoever" (CMG Suppl. 2, pp. 200–201). Galen gives no context for Herophilus' "sophism," but he offers as a parallel (if not an explanation), "Either what is seen and what sees are body, or what is seen is body but what sees is incorporeal, or the converse, or both are incorporeal." By so proceeding, Galen says, one can prove that we do not see at all (pp. 203–204).

I am not satisfied that I know what Herophilus was talking about, but Galen's interpretation seems to make it likely that the subject is "What makes the body work?" That is, when Herophilus talked about causes he was asking the questions Dogmatics asked. His answers would relate, probably, to *pneuma* as the controlling factor in the body: whatever makes the kinetic nerves go, whatever makes the pulse normal or abnormal. While Herophilus investigated the nerves and pulse as he did, then, he likely looked for the most accurate way to express the implications of his findings. If, as seems likely, he experimented in life and death on condemned criminals, I suspect that he was looking for a cause, corporeal or incorporeal, of bodily functions. Galen's words about Herophilus on causes indicate another kind of experimentation, with words, to see whether *aition* can be properly used with other words. His view of his own place in the history of medicine is not stated explicitly in his extant writings, but we can fairly infer that the ancient tradition correctly represented him: he was progressive and dogmatic in a manner appropriate to the period immediately subsequent to Aristotle; he may well have been trying to found a school of medicine on the model of the Academy and Lyceum, since people spoke in his lifetime and immediately after of "Herophilus and his house."[24]

---

24. "Memoirs of Herophilos and His House" was the title of a chatty work by Bacchius, in the immediately succeeding generation (see below, p. 202). Erasistratus, too, had a "House," about which Strato wrote (Galen, K 11.196).

We may say, then, that any notion of Herophilus as proponent of Hippocratism is groundless, as is any notion that Hippocratism was a significant factor in medicine in Herophilus' time.

Similar conclusions are warranted concerning Erasistratus, though his case is somewhat more complicated. Galen not only likes to call Erasistratus half-dogmatic, but he likes to expatiate on the notion that Erasistratus is an enemy of Hippocratism. I need not deal here with all such charges in Galen's writings, since we saw above (p. 78f) that much of Galen's venom was aimed at Martialius and perhaps other contemporary Erasistrateans and that much of Galen's historical argument was rhetorical one-upmanship. It will suffice to deal here in a summary way with what seem to be the major questions about Erasistratus' orientation to what medicine was and what past tradition represented.

I have found no evidence to dispute Fraser's contention that Erasistratus and Erasistrateans were never associated with Alexandria. The story of his diagnosing love as what was wrong with Antiochus is not very compelling evidence that he spent his career in Antioch,[25] yet it is the strongest evidence available. Fraser's further contention, that he did not work in anatomy, seems to go in the face of evidence. Indeed, Galen tells us that Erasistratus did research on the nerves only when he was an old man. But the anonymous treatise on anatomy attributed to Rufus and published by Charles Daremberg and Emile Ruelle appears to compare the views of Erasistratus and Herophilus.[26] Indeed, Erasistratus sounds perhaps more sophisticated: he said that there were two types of nerves, the aesthetic which have their origins in the meninges and are hollow, and the kinetic which originate in the cerebrum and cerebellum. Herophilus spoke of voluntary nerves which originate in the cerebellum and spinal marrow, those which grow from bone to bone and muscle to bone, and those which bind the joints (Dar.-Ru., pp. 184–185).

25. The story is mentioned by many people, including Galen, who attempted to match it in his own medical experience (K 14.630–635). Cf. Fraser's "Career of Erasistratus," pp. 521 ff., esp. 533–534.
26. *Oeuvres de Rufus d'Ephèse* (Paris, 1879; repr. Amsterdam, 1963), cited in the text as Dar.-Ru.

From this at least one would infer that Herophilus' old-fashioned nomenclature was improved by Erasistratus and that in some fashion Erasistratus pursued the question of the source of the nerves.

Such questions are important here only for clues about Erasistratus' orientation to medicine. Two descriptions of predecessors by Erasistratus remain. He praised the therapy of Chrysippus which avoided phlebotomy (but cut off blood from the extremities (K 11.148-149), and he cited Petronas as one extreme in dietetic theory as opposed to Dexippus and Apollonius as the other extreme, perhaps calling them the disciples of Hippocrates. Their extremes, he said, proved that a logical approach to medicine was necessary (CMG 5.9.1, pp. 125-126). My suspicion is that he initiated a new kind of dialogue in medicine, considering his predecessors alogical or prelogical and that he was self-consciously trying to found a sect. Galen is inconsistent in characterizing him—now he is the wise follower of Hippocrates, now he is Hippocrates' brazen, sophistic opponent.

Galen habitually expresses irritation at Erasistratus for ignoring Hippocrates and the concepts that he (Galen) found in Hippocrates. Thus, on digestion: "While Erasistratus, for some reason, refuted some foolish opinions at great length, he entirely passed over Hippocrates' view without so much as deigning to mention it, as he did in his work *On Swallowing*. In that work he did go so far as to mention the term 'attraction,' writing as follows: 'The stomach does not appear to exercise attraction'" (*Natural Faculties* 1.16, *Scr. Min.* 3.145). Galen says that he would be content if Erasistratus had said, "Hippocrates is wrong," but he says nothing of Hippocrates. Similarly, in the case of black bile, Galen is furious that Erasistratus did not discuss the subject nor mention his predecessors to say that they were right or wrong. Galen attributes the failure to jealousy of Hippocrates (CMG 5.4.1.1, pp. 85-91).[27] Galen's other mood in reference to Erasistratus is well exemplified by the work *On Habits* (*peri ethôn, Scr. Min.* 2.9-31), the thesis of which is that

27. For other examples of Galen's irritation at Erasistratus' refusal to mention Hippocrates or other predecessors, see K 2.60-61, 2.133-123, and 11.159.

there is grave medical danger in changing from one's usual habits. Erasistratus and Hippocrates are the "most renowned of physicians," who agree perfectly on all aspects of the subject. Hippocrates is quoted from *Regimen in Acute Diseases, Aphorisms,* and others. Erasistratus is quoted from book 2 of *Paralysis,* a long passage in which he quotes his own work on *General Principles (hoi katholou logoi)* in which he attempted to make clear all one must consider to keep the art from being filled with imperfection (pp. 17–18). Galen also appears to cite Erasistratus' work on *Habits* ("*en tôi peri tôn ethôn logôi,*" p. 28). In none of this material does Erasistratus mention Hippocrates or any other predecessor. When he is polemical it is against error or carelessness.

There is one instance, in Galen's commentary on *Aphorisms,* in which Galen appears to think that he has caught Erasistratus in egregious error on historical matters and slander against Hippocrates. It is instructive of Galen's method and also of Erasistratus' style, since Galen quotes to prove his point:

> We must speak of the disease called lientery. Previous physicians divided this affection into three parts, lientery, dysentery, tenesmos. Some in their distinctions observed and spoke of its differentiation: when the discharge from the intestines is bloody and mucous they called the affection dysentery. If the discharges are undigested but mixed with bloody mucous material, they called it lientery. And when what passes is bilious and mixed with bloody and mucous material they called it tenesmos. [K 18 A.6–9]

Galen responds that he could not imagine how Erasistratus could talk so. No contemporary of Erasistratus said such things as Erasistratus claims (Galen lists, as his examples, Phylotimus, Herophilus, and Eudemus), nor did anyone since his time and no one before him. To prove his point about Erasistratus' predecessors, Galen quotes Diocles and Praxagoras. Then he quotes *Affections* ("whether it is by Hippocrates himself or by Polybus, his student"), and then *Regimen* ("ascribed to Hippocrates himself, but some deprive him of it, some attribute it to Philistion, some to Ariston, some to Pherecydes"). Galen seems right. His quotations show that Erasistratus was misstating the case with reference to "the ancients'" use of the term lientery. But the passage also tells us that this is as near as Galen can come to convicting Erasistratus of "slandering" Hippocrates, or indeed of speaking about him. Under such circumstances, my argument from silence has considerable weight. Erasistratus apparently never did discuss Hippocrates in his works. I would hope now that people will cease to say that he did.[28]

Hence, Erasistratus, like Herophilus, pursued the question of cause in trying to set medicine on a new intellectual basis and tried to do so without sounding naive. Galen found the dogmatism incomplete and in his abusive moods gave Erasistratus a false place in the history of medicine. I shall offer one more example of Erasistratus' insistence on logic, as well as his vague, antithetical style in talking about medical history and his emphasis on the need for a complete theory.

> Most men now and in the past have sought the causes of fevers, and have wanted to inquire of the patients, to learn whether the disease began when the patient was chilled or fatigued or over-filled or some similar cause. But they are not truly or usefully seeking causes of disease. If cold were the cause of fever, the more one was chilled the more feverish he would be, but that does not happen. Some people, who have come into the extremest danger from bitter cold and have been saved from it, do not become feverish. Similarly with fatigue and plethora. Many who have fallen ill with worse fatigue and plethora than when people are feverish yet have recovered from their ailment.

Galen quotes this, in his work on *Precipitating Causes* (CMG Suppl. 2, p. 25), from Erasistratus' work *On Fevers*. It apparently

28. I do not think that I need cite and refute the numerous groundless statements on the subject that have been made, such as the following by Sir T. Clifford Allbutt: "[Erasistratus'] denunciations of Hippocrates, and those of his Roman disciple Asclepiades have loomed, in the clouds of controversy, enormous.... Yet on the other hand we are told that Erasistratus learnt Hippocratic treatises by heart, as he did Homer, and recited them in public" (*Greek Medicine in Rome* [London, 1921; repr. New York, 1970], p. 155). Allbutt's observations, and similar ones, lack evidence, and are misleading.

is the opening of a long argument to the effect that fever comes from putrefaction of blood that has gotten into places where it does not belong (such as arteries). It is reminiscent of the other early dogmatic physicians in its rejection of simplistic reasoning about causes.

In summary, I believe that we can say of all the early dogmatics that they are not followers of Hippocrates nor are they his opponents. They were themselves creating schools, but their writings were in large part similar to the works of the Corpus Hippocraticum in their ignorance of schools of medicine and of any "father of medicine" to whom they might orient themselves. Herophilus or Erasistratus probably would have expressed views of the history of medicine very like those of Menon: "Some say this, some say that, all need improvement; the ancients were cruder than we sophisticated moderns."

Hence my search for the Hippocratic tradition has not yet encountered any responses to Hippocratic doctrine (other than those of Plato and Menon), nor indications of the existence of a Corpus. The generations immediately succeeding Herophilus and Erasistratus began the change in Alexandria.

THE COLLECTION OF THE CORPUS
AND EARLY WORK ON IT

We do have information about the collection of old medical works by the Alexandrian Library. The story begins in the reign of Ptolemy Euergetes I (246–221 B.C.). Zeuxis, the Hippocratic commentator, told it approximately a century later, and Galen transcribed it from him. The excerpt which follows here is from Galen's commentary on *Epidemics* 3. The question at issue is the authority and source of the symbols, Greek letters appended to the case histories apparently as a summary of their contents.

I would be embarrassed to deal in such foolishness if I had not dealt with the whole medical art before coming to the commentaries on Hippocrates' works. Nobody who knows medicine can learn anything from my commentaries besides what is in my medical works, where I wrote systematically enough that even the stupid can understand the meaning of what is said. But they will have a knowledge of history, which makes the multitude admire practitioners because they think that those who have read history and are full of bits of information understand the theoretical basis of the science.

Hence, I will tell the full history of the symbols since my friends and colleagues would like it to be set down once here. What I am going to say was said by Zeuxis in his first comment on the present book. Perhaps it would be better, as I do in such cases, to send those who want the history to read it in that book. But since Zeuxis' commentaries are not in demand, and are scarce, they asked me to relate it, beginning from Mnemon [a mock epic touch by Galen]. Some say that he took the third book of *Epidemics* from the great Alexandrian Library to read, and that before he returned it he added the symbols in it in ink of similar blackness and in letters like those in the manuscript. But some say that he brought the book from Pamphylia with the symbols already inscribed. And, they say, the Ptolemy who was then king of Egypt became so greedy for books that he ordered that the books of everyone who arrived by ship be brought to him. After he had them copied on new paper, he gave the copies to the owners of the books that had been brought to him on the debarkation, and deposited the confiscated books themselves in the library with the inscription "Of Those from the Ships." And they say that the *Third Epidemics* was found to be one of this kind, inscribed "Of Those from the Ships by the Redactor Mnemon of Sidon." But some say that it was not inscribed "by the redactor," but simply with Mnemon's name since the king's assistants inscribed the names of all the travelers on those of their books which were put in storage, because they did not take the books straight to the library, but stored them in some houses in heaps.

That Ptolemy was so eager to acquire books is well witnessed by what they say he did to the Athenians [Galen tells here the story of the piracy of the Athenian state copies of the tragedies.]. Well, then, Mnemon, whether he himself brought the book [to Egypt] or took it from the library and wrote in it, seems to have done so for profit. He said that only he could understand what the symbols meant, and he charged for interpreting them. If that is true, it is more credible that the library had the symbols. He would have been suspected if he himself had brought the copy from home. [CMG 5.10.2.1, pp. 78–80]

Galen's tale from Zeuxis is important, vague as it is, for telling us what the Alexandrians themselves knew of the Corpus. Nothing in the sources suggests that any Coan library was brought en bloc to Alexandria; that modern scholarly myth can be ignored because it would be in the sources if it had any basis at all. Modern scholarship arrived at the conjecture by a logical route, as we have seen: Galen said Hippocrates was dogmatic, as were Praxagoras of Cos and Herophilus. Herophilus studied with Praxagoras, Galen says (K 10.28). When it was imagined that he commented on Hippocratic texts it could also be imagined that he collected them, that he brought a Coan library. Insofar as these hypotheses piled on hypotheses, can be tested, they fail. If the history were as has been conjectured, the acquired library and Herophilus' interest in it would be recorded. But the sources tell of confusion even among people who worked at the library in Alexandria as to whence came the books that were piled up there and later catalogued and attributed to Hippocrates.

The medical books were collected in haphazard fashion. Many were likely the personal copies that had been made for the private use of the travelers from whom they were snatched. It is not surprising, therefore, that they would be anonymous or that some would be composed of material from more than one source. *The Nature of Man*, so important to Galen, is apparently such a composite work, and Galen elsewhere offers his own fantasy about how that book was composed. Probably people could get a good price by selling books to the library, especially if the books bore the names of famous men. Therefore, he imagines, someone took the brief Hippocratic treatise *Nature of Man*, added the nonsensical anatomy of veins, and then sold the whole as Hippocrates' work (CMG 5.9.1, pp. 70–75). Galen's fantasy and his ignorance about the provenance of the Hippocratic books reveal the ignorance of his sources. I think that we cannot pretend to know more than Galen knew about the Corpus Hippocraticum in

Alexandria. But consider one interesting coincidence: when the library sorted out its old medical works and styled them all Hippocratic, it had two works whose first parts were titled *Nature of Man* and whose second part was titled *Regimen in Health.* I have given reason to believe that one, the present *Regimen,* was the work of Hippocrates. The other, the present *Nature of Man,* is the subject of Galen's fantasy and the basis of his Hippocratic elemental theory. Both remained in the Corpus along with other miscellaneous works whose wide differences in quality and point of view could not but be apparent to anyone who analyzed them.

Bacchius is the first person known to have dealt with the works of the Corpus.[29] The later tradition of citations of readings in the Hippocratic works goes no deeper than him as we shall continue to note below. What did he do, with what purpose, and what did he think of the Corpus? I infer the following: Bacchius was a literary man. He wrote a glossary—an explanation of archaic and difficult words in the old medical works, using the *Lexeis* of Aristophanes of Byzantium as a source and offering many quotations from classical poets and prose writers to illustrate the *glossae* (rare words) of Hippocrates (cf. Erotian, N p. 5. lines 1–3, and Galen, K 19.65). He copied, or had someone copy, at least one Hippocratic work, the work we call *Epidemics* 3 (CMG 5.10.2.1, p. 87). And he wrote a book of *Reminiscences of Herophilus and His House,* which gives one anecdote about the physician Callianax: when a patient complained to Callianax, "I shall die," he responded with the line from Homer, "Patroclus is dead, and he was a better man than you."[30] We do not have the context for that anecdote, but the title of Bacchius' book suggests that Bacchius wrote his memoir to record his as-

29. Basic information on Hippocratic glossography before Erotian is well laid out in Max Wellmann's *Hippokratesglossare* (Berlin, 1931). Xenocritus and Callimachus were glossographers near Bacchius' time who may have preceded him. Erotian speaks as though their works were unknown to him (N 4.23–27). Erotian will be cited from *Erotiani Vocum Hippocraticarum collectio,* ed. Ernst Nachmanson (Uppsala, 1918).

30. CMG 5.10.2.2, p. 203. The anecdote was told by Zeuxis to illustrate remarks in *Epidemics* 6 about need for medical decorum and consideration for the patient's emotional state. Galen quotes Zeuxis' remarks.

sociation with a college of Alexandrian medical men grouped around Herophilus—presumably his students and dependents.

Bacchius' glossographical work on the ancient medical writings was, I think, aimed at forwarding cultural appreciation of the Greek language and Greek heritage. People who worked at glossography after him had such purposes and acted as though he had.[31] They used his work and attacked it with some acrimony, but the acrimony is literary, not like that of the sectarian medical disputes to be discussed below. One example of the way a word was treated in the glossographical tradition is worth much description. Here is what Erotian gives for the word *elinyein,* from Epidemics 6 (L 5.268.7):

> *elinyein:* Bacchius in Book 1 says "laziness" (*argein*), lack of occupation (*scholazein*), asserting that the Eleians, for *scholazein,* say *elinyein,* while the Thymbrians say *argein.* However, Heraclides of Tarentum in Book 2 of *Against Bacchius, Concerning Hippocrates' Language,* says that the word is derived from *heile,* the heat and glow of the sun, whence is derived also, they say, *alea,* "warmth." *Aleanthes* is olive oil turned white by the sun. And since people who are warmed by the sun tend to inactivity, *elinyein* is used to mean lack of occupation. I [Erotian] think that Heraclides' etymology is overdone, and that Bacchius is correct.   [N p. 106]

What the medical text, *Epidemics* 6, has to say is "*Elinyein* is not good, but rather exercise." Its meaning is clear, but the obsolete word generated interest and dispute. Unfortunately, modern scholars are as susceptible to anachronism as anyone else. They have assumed that Bacchius must have concerned himself with the same questions that oppress them and must have had views on the Hippocratic Question. (Indeed, the spurious commentary on *Humors* suggested as much.) But that assumption should be discarded, and the effort of imagination should be made to view the early work on the old medical writings in its true light.

31. Erotian's (first century A.D.) statement of purpose can stand as a description of the purposes of previous glossographers: Hippocrates is important, says Erotian, because he is useful for literary instruction. He is useful for physicians especially because in reading him they can learn new things and test the ones they already know (N p. 3, lines 1–10).

Bacchius used a large and varied group of treatises which are in the Corpus, probably including *Prognostic, Prorrhetic, Humors, Epidemics* 1, 2, 3, 5, 6, *Aphorisms, Places in Man, The Surgery, Mochlikon, Joints, Wounds in the Head, Regimen in Acute Diseases, Diseases* 1, *On the Art.* [32] Most probably the list could be extended if the evidence were not so meager. His work seems to have been arranged like Erotian's by the order of the occurrence of the words in the medical works glossed. Epicles abridged the work and arranged the glosses alphabetically in the first century B.C. (cf. Erotian 5.5, N 7.23–8.5). Such an arrangement makes sense in view of the papyrus rolls on which the work had to be read: they preclude thumbing back and forth through an alphabetical list. [33] And possibly later careless references to his work as "commentary" (such as Galen's in the commentary on *Aphorisms,* K 18 A.186) derive from the order of its presentation, despite the title *Lexeis.* When Galen is using good sources and speaking carefully, he distinguishes the glossographer Bacchius from the later commentators on Hippocrates (for example, on *Epidemics* 2, CMG 5.10.1, p. 230).

THE EMPIRICS AND HIPPOCRATIC "DOCTRINE"

The views and the importance of Hippocrates first became a subject of discussion in doctrinal disputes that arose out of the empiric reaction against dogmatic medicine. I wish that I could present a clearer picture of the development of the doctrinal dispute than I am able to do: how it began and how it developed and where the various Empirics figured in it, but only a few

---

32. These I infer from Erotian's citations as Nachmanson assigns them. Wellmann, *Hippokratesglossare,* p. 2, gives a somewhat larger list, including *Airs Waters Places* and *Nature of the Child,* which were glossed in Epicles' epitome of Bacchius.

33. On the development of alphabetization and the uses to which it was put, see Lloyd W. Daly, *Contributions to a History of Alphabetization in Antiquity and the Middle Ages.* Collection Latomus 90 (Brussels, 1967). There were some alphabetic glossaries in the third century B.C.

fragments remain from the early period, ca. 250 to 100 B.C.[34] The first coherent report of the dispute of the Empirics against the Dogmatics comes from Celsus in the first century A.D. But Celsus' report is colored by the contribution of Asclepiades (first century B.C.), the latter-day dogmatic, who launched a counterattack against the Empirics. Galen, as usual, is the source of much of our solid information. Galen offers the dialectic of sect versus sect in various works, very much in Celsus' manner, but with refinements which had been offered by Menodotus, an Empiric of the early second century A.D.[35] Galen also offers bits and peices of information which, with other bits and pieces, provide an outline of the early quarrel. Celsus' report of sect versus sect was written two centuries after the beginnings of the dispute and Galen's works on sects more than three centuries after. Of more concern than refined dialectic of the quarrel between Empiric and Dogmatic as reported in later centuries is the attribution of Empiric doctrine to Hippocrates in the early stages of the quarrel. In some early stage of their attack on dogmatic medicine, the Empirics began to use the Hippocratic Corpus as a standard of nondogmatic medicine, a source of collected experience of observation of symptoms and of methods of treatment that were not controlled by dogmatic theories. And the Empirics also found in the Corpus enough theoretical material to support their position against the Alexandrian Dogmatics whom they were attacking. I shall outline the Empirics' case against the Dogmatics, and then try to assess their use of the Hippocratic Corpus.

34. Fragments and other material relating to the Empirics were collected by Karl Deichgräber, *Die griechische Empirikerschule.*
35. Galen's remarks in *On Medical Experience,* ed. and trans. Richard Walzer (Oxford, 1944), p. 87, indicate that Asclepiades wrote on sects in the form of an outline of the theoretical position of Serapion followed by answers to Serapion. (Celsus seems to draw on that work, but to soften Asclepiades' judgments.) Menodotus responded in kind to Asclepiades' writings, and Galen drew on Menodotus' outline of sects. But Galen responded more favorably to Menodotus' Empiric view in his early than in his later writings (see above, p. 76). Who provided the particular outline of Methodism used by Galen (and perhaps by Menodotus before him) is not clear to me.

Certainly the Empirics attacked the Dogmatics (they may have given them the name) for practicing bad medicine: for murdering patients, in effect, in the name of their logic. *Logos,* always an ambiguous and inclusive term, had the meaning of rhetoric as well as logic: *logiatroi* were physicians who talked much and did nothing, who worked out logical theories in their studies or laboratories, and who spoke well in public displays of learning, but who never saw a patient, understood a disease, or cured anything. Galen reflects the Empirics' accusation when he is more than usually scurrilous against Erasistratus: "Maybe, as they say, you had contempt for seeing sick people, and stayed at home to write up your thoughts" (K 11.159), and Galen uses the term *logiatroi* whenever he needs a word to describe healers in theory but not in fact.[36] The terminology of the quarrel and the quarrel itself, however, probably developed after the time of Herophilus and Erasistratus, after the creative period, when their followers were consolidating early gains and turning them into dogmata. So the nearest contemporary evidence would suggest. The nearest contemporary evidence is particularly good because it is virtually irrelevant: it is a half-apt analogy between armchair historians and armchair physicians which the historian Polybius, writing between 150 and 120 B.C., adduced to prove that a man of affairs such as himself is a superior historian. I quote him in part:

> The logical part of medicine which came mostly out of Alexandria, from the Herophileans and Callimacheans, as they are called there, comprehends one element of medicine [the others are

36. The word *logiatroi* was coined by the Empirics, I infer, as a term of abuse of the dogmatics. In another attack on the dogmatics, which seems to come from the empirics, Galen says, "People who deal with the facts are aware of this, unlike the sophists of Alexandria whose books are full of long and false stories for boys fresh from the farm. They never saw a patient, nor bothered with such [symptoms of disease]. Dietetics was entirely neglected by them" (CMG 5.10.2.2, p. 10). The best direct example I have found of the Empirics' style of argumentative abuse is Heraclides' comparison between Andreas and Pamphilos' descriptions of medicinal herbs and heralds' descriptions of a runaway slave: "They get their information from those who know and use it as an incantation, but they would not know the slave if they were standing beside him" (K 11.796).

dietetics and surgery and pharmacy], but by its claims and preten-
sions it makes people imagine that the others are useless. When
you confront such thinkers with the facts, by offering them a
patient, they are as helpless as someone who has never read a case
history. Patients who were not very ill have entrusted themselves
to such physicians by virtue of their reputation [or "because of the
force of their rhetoric," the word *logos* again] and have en-
dangered their whole existence.[37]

Polybius enjoys an attack on the pretentious intellectual who
does not know his business. The particular attack he here
paraphrases was, I think, formulated by the Empiric physicians,
who, like all reformers, denigrated their predecessors, upon
whose heritage they wanted to make improvements. Polybius'
analogy assures us that such medical denigration and quarreling
occurred within his lifetime. The details of the Empirics' attack
on dogmatic medicine, the improvements they urged, and the
way they used Hippocrates must be inferred from later reports.

The Empirics rejected the search into causes of health and
disease by saying that immediate causes were obvious and that
hidden causes were not discoverable. Conjectures about the
mechanism of breathing or the source of the pulse or the manner in which food is digested are not relevant to medical practice. The question in medicine is: How does one treat a sick
human being? Their answer was: One gets a notion of his affection by noting his symptoms, and one treats him with the remedies that experience has shown effective for such affections.
That is what medical practice is and if one pretends that it is
anything more, he is a fraud (Celsus, *Proem.* 27–32). The quarrel
can be considered an example of the eternal conflict between
laboratory or theoretical medicine and clinical or practical
medicine; perhaps it is, but the too easy analogy can be misleading. In any case, the Empirics attacked the medicine of the
Dogmatics in the name of better medical practice, undoubtedly
with great sincerity. They attacked the followers of the dogmatic
group: a genuine "school" had come into existence as logical

37. Polybius, ed. Theodor Büttner-Wobst (Leipzig, 1882–1905), 12.25d4–5.

medicine passed its creative period, and the Empirics intended to replace it with their own school.

The Empirics attacked the study of anatomy by dissection and vivisection: the corpse and the body distorted by pain do not give insight into causes of health and disease or into useful treatment (Celsus, *Proem.* 40-44). Against the authority of the *logos*, the Empirics provided their own epistemological theory, giving authority to *experience:* to the physician's own experience and to what he learned from research (*historia*) into the medical experience of others he could trust.[38] Here we reach the crux of our subject, their use of the Corpus Hippocraticum.

The Empirics gave authority to the Corpus Hippocraticum and out of it they created a Hippocrates in their own image, whom they opposed to the practitioners of the dogmatic school. They found in the works of the Corpus the kinds of descriptions of the phenomena of disease, surrounding circumstance, and effective treatment that fitted their own patterns of research.[39] They found, particularly in the *Epidemics,* the repeated injunction that this and that factor must be investigated to determine relevance (*skepteon* and *zeteteon* are the common terms).[40] In the *Aphorisms, Prorrhetic,* and other works of condensed experience, they found descriptions of the precise order of symptoms in diseases and juxtapositions of symptoms with environmental and other factors in precisely the manner they thought proper.[41] They were pleased to find that Hippocrates preferred observation to theory, that he emphasized phenomena that theory would not have prepared one to expect (Galen in CMG 5.10.2.2,

38. For *historia* as the Empirics defined it, see Galen, *On Sects for Beginners, Scr. Min.* 3.3, and Deichgräber's summary, *Empirikerschule,* pp. 298-301.

39. The Empirics say they do not know where Hippocrates and others found their methods of treatment (that is, there is no deductive process that would produce them), but the important thing is to use them well (Galen, K 8.142). They considered the Hippocratic prognosis of critical days to be a method derivable only from experience (*On Medical Experience,* ch. 9, and Galen, *On Critical Days* K 9.774).

40. For Heraclides' views on *zeteteon* as a byword of Hippocrates, see *CMG* 5.10.2.1, pp. 87-88. For similar views by Glaucias, CMG 5.10.2.2, p. 451.

41. For their treatment of the *Aphorisms* as an empirically oriented work, see Deichgräber, *Empirikerschule,* Frags. 362 and 363 (from Galen, K 17 A.24 and 18 A.345-354). See also Frags. 331, 332A for *Prorrhetic,* 327 for *Prognostic,* 348-349 for *Epidemics* 4 and 5.

p. 174). Furthermore, they could find approximations of their own theoretical position in the works of the Corpus. Beyond the general injunctions to observe and investigate, they found descriptions of what medicine is that resembled their own. Celsus' report of their views shows that the treatise *Ancient Medicine* was particularly congenial to them. *Ancient Medicine* offers an anthropology to support its view of what the Art of medicine is and where it came from: the Art (*techne*) originated in people's experience of the effects of diet and environment on their health; raw and harsh food had bad effects, as did sudden change, and their opposites had good effects; men differed from animals in their needs. *Ancient Medicine* infers that the systematic collection and use of such information was the origin of the ancient medical art and proceeds to argue that thence comes a paradigm of all medical research and reasoning, while philosophical hypotheses about hot, cold, wet, and dry are useless.[42] Celsus' report of the Empirics' presentation of their views reads like an excerpt from *Ancient Medicine:*

> Even in its beginnings, they [the Empirics] add, the art of medicine was not deduced from such questionings, but from experience; for of the sick who were without doctors, some in the first days of illness, longing for food, took it forthwith; others owing to distaste abstained. . . . When this and the like happened day after day, careful men noted what generally answered the better, and then began to prescribe for their patients. Thus sprang up the art of medicine. . . . It was afterwards, they proceed, when the remedies had already been discovered, that men began to discuss the reasons for them.[43]

Another piece of evidence shows that, in their interpretation of Hippocrates by Hippocrates, Empirics interpreted the dark and difficult passages in the *Epidemics* in the light of *Ancient Medicine:* "They say that the words of Hippocrates in this passage are the same as those found in *Ancient Medicine,* and the author has here

42. *Ancient Medicine,* chs. 4–8. The Empiric comparison between learning medicine and learning to pilot a ship (Celsus, *Proem.* 31–32) seems to allude to the same comparison in *Ancient Medicine,* ch. 9.
43. Celsus, *Proemium* 33–36; Celsus' own (dogmatic) view of the origins of medicine is that medicine came out of philosophy (*Proem.* 5–8).

only shortened and confirmed them: specifically, one cannot, according to his words, take hot, cold, wet, and dry as bases in the healing of diseases."[44] Galen is, of course, offended by that view of Hippocratic science. He gives us the information in one of his expansive moments in which he is apparently transmitting information given by Rufus of Ephesus. Galen precedes his unusual burst of information about those who have interpreted the passage in accord with *Ancient Medicine* by saying, "In my view it makes no difference if I also mention those who have interpreted this passage in a way contrary to Hippocrates' opinion."

There is an interesting historical irony here. Littré did not know Galen's statement about *Ancient Medicine*. Yet in his own reconstruction of Hippocratic science (purging away Galen's prejudices and putting in his own) Littré made *Ancient Medicine* the prime Hippocratic work by which the rest of the canon was to be tested (1.294–320). In so doing, Littré was in part restoring the Empiric view of Hippocrates which began our Hippocratic tradition. But that tradition began in the Hellenistic period, not in the Classical period where Littré would put it. Despite Galen's usual reticence about views of Hippocrates different from his own and despite his rhetorical posture according to which the history of medicine is a series of acceptances or rejections of Hippocratic dogmatism, his evidence appears clearly to point to the Empirics as the first interpreters of Hippocratic science. The tradition of study of the text goes slightly deeper, to Bacchius. But Bacchius attributed no doctrine to Hippocrates, while the Empirics clearly did. What each individual Empiric contributed cannot, I think, be known, as our evidence stands. The commentaries of Heraclides and Zeuxis were the sources of information for later antiquity. They were the first commentators on the Hippocratic Corpus, apparently, and they were said to have commented on "all" the works, whatever that means.[45] Glaucias,

---

44. Galen on *Epidemics* 2, CMG 5.10.1, p. 220, immediately preceding the statement that Heraclides had found a reading for the passage in an old manuscript.

45. Galen repeatedly says that they commented on all the works, but he himself probably did not know what his statement meant. I have been unable to compile a list of works they commented on.

apparently an Empiric glossographer, preceded them, and Zeuxis later loved to criticize Glaucias for failing to make sense of passages according to their whole context (cf. CMG 5.10.2.2, pp. 113 and 217). Two empirics named Apollonius, a father and son, wrote voluminously on the symbols in *Epidemics* 3 and adduced the verbal habits in other Hippocratic works to support their arguments, which were directed against a Herophilean, Zeno (CMG 5.10.2.1, pp. 86-87; Zeuxis is the source of the information). Philinus and Serapion, Empiric theoreticians, may have talked about Hippocratic science.[46] A good sample of Heraclides' style of commentary and of his manner of drawing on predecessors appears in Galen's commentary on *Epidemics* 2. It is a medically insignificant passage on which Heraclides made an elegant textual emendation: instead of the lady's tail pointing to sex, her house door opened toward the temple of Aphrodite.

> Some said that Hippocrates in these words used metaphor by way of analogy. He would indicate the wish of the woman for copulation and uses an allusive expression to beautify the sense with the words "the tail (*ourai*) bends toward sex (*aphrodisia*)." He alludes to the fact that creatures who urinate from behind, i.e., females of all species (and the male, so Aristotle says, of the rabbit, the ape called lynx, and the camel), move their tail when they are sexually aroused and rub their sexual parts with it, since it is near the sexual organs. They say, then, that Hippocrates had this in mind when he said of this woman that her tail bends to sexual pleasure. . . . Some said that copulation by these animals results from the male placing himself on the female. . . . Andreas said that they mount as a rooster mounts a hen. . . . We see that in copulation their tail must go upright because of its nearness to the sexual parts. In what Hippocrates says, "The tail was bent for sexual pleasure," he merely indicated the irritation toward sexual intercourse that affected this woman. This explanation is far from the

---

46. Philinus wrote six books of lexicography "against Bacchius' *Lexeis*" (Erotian, N 5.4. Erotian records his definitions of three words.). Galen ironically calls Serapion "new Asclepius" because he criticized Hippocrates and claimed that he himself was the first truly nondogmatic physician (Deichgräber, *Empirikerschule*, p. 86). Serapion's view on a passage in *Epidemics* 6 was given by Rufus, whom Galen quotes (*CMG* 5.10.2.2, p. 411), but in what connection Serapion expressed his views is unclear.

mark, in my view. . . . [Another explanation is that] Hippocrates describes by his words the protrusion of the uterus: *aphrodision* means genitals and "tail" the mouth of the uterus, as we often call the penis "tail." But I myself am far from accepting that explanation. . . . In order to be complete, we must, besides what we have mentioned, say how Bacchius understood it: in the second of his books explaining Hippocrates' expressions he comes to speak of this word. He says "Hippocrates, by the 'tail bends toward the aphrodision' meant that this woman at that time leaned toward copulation." I do not see how this interpretation relates to Hippocrates' words, since there is nothing there that points to this express meaning.   [CMG 5.10.1, pp. 232–233]

Heraclides' elegant emendation followed this rather tedious description of his predecessors' views. Unfortunately, he is as vague about individual predecessors as Galen generally is, but it is clear that his tradition goes back only to Bacchius, and no further.

Our earliest extant commentary on a Hippocratic work comes from first-century B.C. Alexandria, written by Apollonius of Citium (who is to be distinguished from the Apollonii, father and son, mentioned above).[47] The work is less a commentary than an illustrated appreciation of the surgical work *Joints*: a transcript and paraphrase of the work's mechanical procedures for setting dislocations, interspersed with illustrations and with brief explanations and comments on the excellence of the work. Apollonius is useful because he shows what Hippocratic medicine was conceived to be in first-century B.C. Alexandria, after the Empirics had set the Hippocratic tradition going. To the Ptolemy for whom he writes, Apollonius exhibits his credentials: he has studied with Zopyrus at Alexandria, and Zopyrus practices medicine precisely as the "most divine" Hippocrates did (10.5, p. 12.2–4). Later in his work, when he is talking about the setting of the thigh bone, Apollonius quotes Hegetor, a Herophilean, who has "embraced the much touted anatomy" and who speaks of the futility of the method "now in vogue" for

47. Apollonius' work is available in CMG 11.1.1 (1965). I cite the work according to page and line of that edition.

setting the thigh. I translate the passage, which is not available in English, as far as I know, to give the flavor of Apollonius and of Hegetor, whom he quotes:

> I am amazed at those Herophileans who have embraced the much touted anatomy, and especially at Hegetor. In his *On Causes* this is how he talks of the dislocation of the thigh bone and explains what pertains to it: "Why don't they try to find some other way to set the head of the thigh bone besides those now in vogue so that when it goes out of joint and is reset, it will stay? They only depend on habit (*tribê*) and get their theory from the analogies of the lower jaw and the head of the humerus, and ankle and knee and fingers and virtually all the joints which are often dislocated. They cannot figure out why this joint only, when it is dislocated and set again, will not stay. If they use what happens for the most part in other joints, they will have no credible reason for thinking that there might be a better way to set the joint, whereby it would stay in place and act as the others mostly do. But if they had realized the cause, from anatomy, as follows: that there is a ligament growing out of the head of the thigh bone that grows into the middle of the socket; when it is there, it is impossible for the thigh bone to be dislocated, but when it is torn apart it cannot be joined again. If it is not joined together, the joint cannot stay again in its place; thus, when the cause is clarified, they will abstain entirely from setting a dislocated thigh bone, and not pursue impossible operations."

Hegetor has not only erred in this, but has led friends of medicine astray with him. He has in no way defeated what Hippocrates says in *Joints,* but by his own inconsistency he makes a foolish proposal in the foregoing. To be brief, I will make my observations to him summary.

Those who depend on habit itself alone, and stay by what is observed by experience (*empeiros*), are not going to agree that generally the dislocated thigh bone which has been set will go out of joint again. Nor will they fail to set it again, if not successful with one. But if what he contends were true, those who use observation would not do as they do now, but in the same way as they proceeded from observations in the case of the joints generally, so in the case of the thigh bone it is likely they would have comprehended peculiar results in the case of the thigh. As a result they are not devising and seeking a better method of setting by reasoning, but they stay with what they have seen empirically. And, that the thigh bone, dislocated and set, is necessarily dislocated again, neither events nor the *historia* of the ancients comprehend. And if anyone ever took care with the subject of joints, Hippocrates did. He, in his concern for truth, though he illuminated many peculiarities of the rest of the joints, did not reveal of the thigh that it cannot be reduced, but on the contrary poured his spirit on the setting of the thigh bone so as to make it a practical subject, and he laid out for each kind of dislocation that is not set the lameness that ensues. [CMG 11.1.1, pp. 78–80]

Apollonius follows this passage with citations of many passages from *Joints* that show the modesty and precision of Hippocrates and express confidence that dislocations of the thigh can be reduced (pp. 82–94). One will note that while Hegetor does not in fact mention Hippocrates, Apollonius defends him anyway.

Did Apollonius, with his "Hippocratic medicine," belong to a sect? Scholars have inferred that he was an Empiric, but he does not say so himself.[48] He does use Empiric sounding language, but only in the single passage I have quoted in which he opposes the mistaken dogmatism of Hegetor. In the remainder of his work, his general laudation of Hippocrates tells us nothing much. I, myself, suspect that he styled himself a Hippocratean, and my suspicion is based on this: we know that he wrote in opposition to the Empiric Heraclides, in eighteen books, and in opposition to Bacchius in three (Erotian, N 5.8–9) on Hippocrates' language. What his commentary reveals of his disputatiousness about Hippocrates' language seems to show a peculiar possessiveness with regard to Hippocrates. Of the word *ambê*, a rim on an instrument for reducing shoulder dislocations, he says that Bacchius did not understand the word, despite the tes-

48. Apollonius' only medical work of which we have knowledge, a work on epilepsy, betrays nothing as to his "school" (see Deichgräber, *Empirikerschule,* frags. 278–280). The same can be said of the few recipes preserved from his teacher Zopyrus (Deichgräber, frags. 267–274). Apollonius adduces only an otherwise unknown physician named Posidonius as an additional witness to Zopyrus' Hippocratism (p. 12). Fridolf Kudlien has offered what seems to me an unlikely and poorly supported conjecture that the man in question is the Stoic philosopher ("Posidonius und die Aerzteschule der Pneumatiker," *Hermes* 90 [1962], 427–429).

timonies he adduced from classical authors. Had Bacchius known what the citizens of Cos called the rungs of ladders he would have been on his way to a solution (28.1–15). Bacchius failed to understand another phrase because of his lack of practical experience in medicine, so Apollonius tells us (16.3–10). Unfortunately, Apollonius' objections to Empiric interpretations of Hippocrates' language are unknown. Apollonius' work shows that "Hippocratic medicine" was not dogmatic in his time. Apollonius may have been an Empiric, but more likely he stepped forth as a Hippocratean.

Apollonius' possessiveness of Hippocrates in competition with other interpreters and his interest in the peculiar "Coan" qualities of the language along with his idolization of Hippocrates the "lover of truth" suggests that we have entered a new stage of the tradition. It is, however, noteworthy that Apollonius does not lay claim to extensive biographical information, finds no miracles in Hippocrates' life, and does not claim philosophical acuteness for Hippocrates. Galen's idealized Hippocrates, in *The Best Physician is Also a Philosopher,* obviously is much more influenced by the letters, *Speech from the Altar,* etc., which are contained in the Hippocratic Corpus than Apollonius was. This difference may give us some purchase on the chronological sequence of the development of the mythological core of the later Hippocratic tradition.

THE MYTH

A series of letters and speeches that became part of the Hippocratic Corpus (published in volume 9 of Littré's edition) relates Hippocrates' life and deeds and places Hippocrates and his sons and students in the history of Cos and of Greece generally. They tell of an (apparently wholly mythical) Athenian plan for an expedition to attack Cos, which was averted by appeals by Hippocrates and his son in the Athenian assembly. They speak of Thessalus' service to the Athenian army at Syracuse and of the marvelous service of Hippocrates, his sons, and his students in curing the Athenian plague and also another apparently wholly mythical plague that affected all Greece. They speak of Hippocrates' loyalty to the Hellenes and his refusal to serve barbarian royalty for any amount of gold. Also they tell of a gold crown and other honors given to Hippocrates by Athens. They tell of Hippocrates' meeting and correspondence with Democritus, the laughing philosopher, whom the world thought mad because he did not accept its values, but who taught Hippocrates to share his view of the world and its ways.[49]

With these works, which I shall call pseudepigrapha (writings with false superscriptions) for want of a better term, begins the circulation of the myths of the idealized Hippocrates whom Galen took for granted. Scholars have inferred that at least some of the material stems from Cos, since it appears to show acquaintance with Coan geography and history and is associated with notions of saving Cos from more powerful states.[50] Whether we can date the thematic material more precisely seems dubious. For example, is it impossible or even unlikely that a work would present Hippocrates as siding with Greek against barbarians if it was written while Cos was under Ptolemaic control? I think not, but Edelstein accepted that view.

Some archaeologists have attempted to assert that there are material grounds for crediting legends about Hippocrates' life. For example, there is a fourth-century inscription discovered at Delphi and published by Jean Bousquet that instructs the members of the *koinon* (commonality) of Coan and Cnidian Asclepiads as to how they should identify themselves with an oath to claim the privileges accorded at Delphi to "Asclepiads by male descent."[51] Because the phrase "Asclepiads by male de-

---

49. The pseudepigrapha are not consistent with one another, nor are they all of a period. Fridolf Kudlien argues plausibly that the first two letters are from the second century A.D.: *RE* Suppl. 10 (1965), 473–474. The Democritus letters are generally dated to the first century B.C., the *Decree, Speech from the Altar*, etc., somewhat earlier. The most extensive study of the letters was done by Robert Philippson, "Verfasser und Abfassungszeit der sogennanten Hippokratesbriefe," *Rheinisches Museum* 77 (1928), 298–328.

50. Cf. Ludwig Edelstein, *RE* Suppl. 6, 1301.

51. Jean Bousquet, "Inscriptions de Delphes," *Bulletin de Correspondance Hellenique* 80 (1956), 579–593. Important supportive material for Bousquet's interpretations was published by Heinrich Pomtow, "Delphische Neufunde, III. Hippokrates und die Asklepiaden in Delphi," *Klio* 15 (1918), 303–338.

scent" is used also in one of the pseudepigrapha (the *Embassy*, L 9.416) and because the phrase "by male descent" is used elsewhere on Coan inscriptions to describe priests' qualifications but is otherwise rare, Jean Bousquet would associate the *Embassy* with a Coan author. Further, because the inscription makes it clear that Asclepiads had privileges at Delphi and because the *Embassy* speaks of renewal of Coan privileges in the presence of Hippocrates and Thessalus, Bousquet seems inclined to credit, at least in part, the historicity of the *Embassy*.[52] But let us pause for a moment to see where such reasoning leads. It would be possible to credit, perhaps, the story that the Oracle sent the Amphictyons to Cos for "the son of a deer and gold" to win the First Sacred War and that Hippocrates' ancestor set them straight as to its meaning: "I am Fawn (*nebros*), this is my son Gold. I will be your general" (L 9.410). We can even, perhaps, credit the manner in which Nebros, a physician, won the war—he poisoned the water supply of the Criseans and gave them intestinal problems that weakened them (9.412).[53] However, that is only the first good deed that Thessalus details in the *Embassy*. Later ones get closer and closer in time to the supposed date of the speech and, as they do, it becomes more and more difficult to credit them: Cos refused to join the Persians, Artemisia besieged them, but gods defeated her fleet; Cadmus, another ancestor of Hippocrates, migrated to Sicily and urged Gelon not to join barbarians against Greeks (9.416).[54] Still later,

52. The dating of the *Embassy* near in time to the Delphian inscription by the Asclepiads is crucial to Bousquet's argument and to Pomtow's construction: if the *Embassy* is late Hellenistic in origin, that is, after the Alexandrian Library's collection of "Hippocratic" material, as is generally accepted to be the case for the *Letters,* then the *Embassy* cannot help us to interpret the fifth- and fourth-century inscriptions at Delphi (Cos has offered very little).

53. As Don Lateiner drew to my attention, the traditions about the First Sacred War involve concern for maintenance of water supply: the Amphictyons took an oath not to cut off one another's water (Aeschines, *On The Embassy* 115). Solon is said to have poisoned the water in another account of the First Sacred War (Pausanias 37.6).

54. Herodotus (7.163–164) tells of Cadmus' abdication of supreme power in Cos, of his migration to southern Italy, and also of his mission for Gelon to take money to Delphi to give to Xerxes if the Persians beat the Greeks. Does our *Presbeutikos* represent a separate tradition or is it offered as a footnote to Herodotus? I cannot tell.

Hippocrates refused to aid the European barbarians when they were struck by plague, but sent his sons and students through Greece to instruct each state how to prevent the plague, which they all did successfully (9.418–420). Some small concession is offered to ancient medical theory, at least, when Thessalus notes that the prevention of the plague differed in each state because the winds, heat, and other conditions differed (9.418). Otherwise one can say that the myth is offered without a realistic sense of what medical practice was like. Finally, Thessalus claims that when the Athenians sent Alcibiades to Sicily, Hippocrates volunteered to send Thessalus to take care of the army—without pay (9.422). Whence, concludes Thessalus, the Athenians should be grateful and not enslave Cos. The putative historical occasion for the *Embassy* is not clear, though Thessalus says the plague was nine years previous (9.420). Since Littré's eloquent treatment of the subject there has been general agreement that the *Embassy* is fantasy.[55] But the fact that the historical situation nearest in time to the writing should be so much vaguer than that of the early sixth century and the Sacred War indicates to me that there is no reason to date the writing of the speech close to the supposed date of its delivery. Hippocrates and Athens are brought into conjunction by the *Embassy*. Political events are manufactured to suit the purpose, but are manufactured in such vague form that they suit Athenian imperialism simply: "Athens should not enslave us. Absolute power is bad." Certainly Hippocrates' prominence is implied by the creation of the myth, but the "facts" of Hippocrates' life in the myth seem in no way substantiated by the *Embassy*.

In sum, while Bousquet would like to use the combination of the inscription and the *Embassy* to prove that privileges were accorded Asclepiads for help to Delphi in the First Sacred War and that those privileges were renewed in the lifetime of Hippocrates, to do so involves considerable suspension of disbelief. All that can be firmly established is that the privileges for

55. Littré discussed the pseudepigrapha in volume 1 of his edition and later returned to the subject when he had suitable opponents for his eloquence who had asserted the historicity, if not the genuineness, of the material in the *Presbeutikos* and other works (vol. 7, pp. V–L).

Asclepiads which the *Embassy* speaks of did exist. It seems clear that the ancients treated such biographical fantasies much as moderns do: obviously, the stories are not entirely true, but maybe they have something behind them. Thus antiquity generally ignored the *Embassy*'s fantasy about the connection of Hippocrates with Athens and the obviously made-up plagues (Pliny, *Natural History* 7.123, is exceptional in this regard as O. Temkin points out to me). The "something" behind the story that the ancients seem to have accepted were the names of Hippocrates' children and students, his general patriotism, and his itinerant life, the things that Galen makes use of. What evidence supports such inferences? Only their accord with the Hippocratic Corpus, particularly with the apparent itinerant practices of the author or authors of *Epidemics*. But, we must ask, isn't it likely that the pseudepigrapha concocted their biography out of inferences from the Corpus? The least bit of skepticism appears to me to lead to the judgment that the pseudepigrapha are not dependable sources of historical information. Because that is my view, I am left with the need to evaluate that most basic bit of fact, Hippocrates' genealogy. This subject cannot be avoided because it is closely related to attributions of authorship of works of the Corpus. I incline to the view, first, and most important, that there is no ascertainable relation between the Corpus and the sons, son-in-law, grandsons, and father and grandfather of Hippocrates, and, second, that the genealogy itself seems to have no authority that can be ascertained—we cannot claim to know that Hippocrates had a son-in-law or who his sons were. While this latter view may appear to be born of excessive and unnecessary skepticism, it seems to be the most defensible approach to determination of what happened to the Corpus Hippocraticum in antiquity.

Polybus, the son-in-law, apparently has the best claim to identity as author, but the claims look very problematic when examined. The claims to authorship as laid out by Hermann Grensemann[56] are of different sorts for *Nature of Man* and for *Eight*

56. Hermann Grensemann, "Polybus als Verfasser hippokratischer Schriften," *Abhandlungen der Akademie der Wissenschaften und der Literatur*, Mainz, geistes- und sozialwissenschaftlichen Klasse, no. 2 (1968), 1–18. Karl Deichgräber,

Month Child and *Nature of the Child.* Aristotle quotes the description of blood vessels in *Nature of Man,* chapter 11, as the theory of Polybus or Polybius. He compares its erroneous notions with the erroneus notions of Syennesis of Cyprus and Diogenes of Apollonia. Why, one would like to know, is Polybus not given a home state?[57] Polybus is cited in the Doxography of Menon, chapter 19, as author of a theory of nature of man and disease that seems very near to that of *Nature of Man,* chapter 3. The papyrus of Anonymus Londinensis is especially tattered at chapter 19. Only the first part (about half) of each line of text is legible. Diels's reconstruction of the missing text was based on *Nature of Man,* and his results have since been used to prove that *Nature of Man* is the source of Menon's report of Polybus.[58] These circular arguments should probably arouse our suspicion, but if we repress it we can say that we have ancient evidence that two parts of *Nature of Man* were attributed to Polybus. This still leaves the question, Who is Polybus? The pseudepigrapha, which give the most vivid picture of Polybus as Hippocrates' son-in-law, do not know him as the author of *Nature of Man,* but only as an actor in the heroic actions against the plague. When we ask when the connection was made between Polybus of the pseudepigrapha and the author of *Nature of Man,* we will find no definite information, but we shall see below that the first people who seem to have made conjectures about authorship of treatises of the Corpus by sons and son-in-law were Dioscurides and Capiton, who put together their scholarly doubts about works of the Corpus with the "information" of the pseudepigrapha. It may be that the coincidence of names between Hippocrates' son-in-

Die Epidemien, pp. 165-166, sets forth the basis for the ascription. Jacques Jouanna, in CMG 1.1.3, pp. 55-59, points out the weakness of the ancient testimony for ascription of *Nature of Man;* cf. his article in *Revue des Etudes Grecques* 82 (1969), 552-562, "Le médecin Polybe. Est-il l'auteur de plusieurs ouvrages de la collection hippocratique?"
57. Aristotle, *Historia Animalium,* ed. and trans. A. L. Peck, Loeb Classical Library (1965), Bk. 3, 511 b22-513 a8.
58. W. H. S. Jones, *The Medical Writings of Anonymus Londinensis,* pp. 75-77, accepts the identification without doubt. So does Grensemann, "Polybus," pp. 57-58, who, like Jones, prints Diels's conjectural text of Anonymus without indicating how much of it is conjecture.

law and the author of *Nature of Man* was the source of their conjectures. Before them, save for the notices in the school of Aristotle, *Nature of Man* was considered to be by Hippocrates, in the same way as the rest of the Corpus.

To return, now, to the starting point of this discussion of pseudepigrapha, Apollonius of Citium is somewhat possessive about the "most divine" Hippocrates, enamored of his personality as "lover of truth" and of his Coan dialect.[59] But Apollonius does not even hint at knowledge of the contents of the pseudepigrapha. If he knew them he appears not to have taken them seriously. The pseudepigrapha seem to be Coan in origin, the product of puffery of the island's past. There is discontinuity in the archaeological record on Cos because the excavations have been at the Hellenistic site. In 366 B.C. the people of the island Cos moved their capital city from the far end of the island to the end near the mainland and the commercial routes.[60] Afterward they built a large temple of Asclepius. By the third century there appears to have been a thriving medical school that exported physicians on request to the Mediterranean area. Inscriptions honoring such physicians indicate Coan pride in their humane service and in their reputation.[61] It is no surprise that Coans laid claim to a continuous tradition which went back to the Trojan War and to the original Asclepiads, Asclepius' sons. Nor would it be a surprise that they composed genealogies to support their

59. Apollonius, CMG 11.1.1, note especially page 10, line 5, 82.4–7, 28:1–15.
60. See G. E. Bean and I. M. Cook, "A Walking Tour," in the *Annual of the British School at Athens* 52 (1957), 119–127.
61. For examples, see Wilhelm Dittenberger, *Sylloge inscriptionum Graecarum* (Berlin, 1915–1924), nos. 528 and 943. Numbers 804, 805, and 806 show that the tradition continued into Roman imperial times. Claudius' generous remission of taxes for Cos (Tacitus, *Annals* 12.61) coincides with renewed literary interest in Hippocrates in Rome at the time Erotian's dictionary was made and Celsus wrote his work on medicine. Apparently, a counter-Hippocratic mythmaking tradition promoted stories that Hippocrates burned the library at Cos (so as to reduce competition with the Hippocratic Corpus?) and had to flee (Tzetzes, *Chiliades* 7.963–967) or that he burned the library at Cnidus ("Soranus," *Life of Hippocrates,* CMG 4.175). Varro's version was that Hippocrates copied the god's prescriptions as recorded on patients' votive tablets, kept the information for himself, and started clinical medicine when the temple burned (Pliny, *Natural History* 29.4).

stories. It is not surprising that they heroized Hippocrates, but it is striking that the heroized Hippocrates is ignored by medical men. Erotian lists the *Presbeutikos* and the *Epibomios logos* as presenting Hippocrates more as a patriot than a medicine man (Erotian N p. 9). Whether he was at all skeptical is not clear to me. After Erotian the pseudepigrapha were in the tradition, but accorded only selective credence such as turns up in conjectures about authorship of the Corpus and in Galen's use of Hippocrates as the ideally unselfish doctor, and also, as we shall see, Hippocrates' apostleship to Democritus which Celsus repeats.

GREEK MEDICINE IN ROME

Creativity in medicine shifted from Alexandria to Rome in the first century B.C. and so did scholarship. Once Hippocrates was injected, by the Empirics, into the arena of medical dispute, he had supporters and detractors. But the more serious dispute was "which side was he on?" whether dogmatic or Empiric. How he switched sides between the Empirics and Galen will be a subject of inquiry, along with the view of medicine's past exhibited by physicians and scholars in Rome in the late Republic and early empire. Asclepiades of Bythinia, who worked in Rome roughly contemporary with Apollonius' Hippocratic medicine in Alexandria, concerns us first. Asclepiades is an elusive figure in various respects. He was very influential, but he aroused great distaste in many people. Eloquent, apparently flamboyant, a controversialist, but also a serious thinker, he declared himself a dogmatic and presented a philosophically based medical system, which, according to Celsus, was the first advance in medicine after the Empirics (*Proem.* 11). The total loss of his writings is unfortunate. However, I think we can use surviving reports of him to arrive at some estimate of his relation to predecessors and his use of the Hippocratic tradition.

If we can believe Galen, Asclepiades overturned all previous dogmas and spared none of his predecessors, including Hippocrates. He called the medicine of the ancients a ministration of (or study in) death (K 11.163). He considered the Empirics' observation without theory nonsensical, and he considered the theories of his dogmatic predecessors wrong or inadequate. He attempted to rationalize medicine by means of a physiology that described all phenomena of disease and health in terms of influx and efflux of material: the body is composed of atoms of various sizes that constantly come and go and that move within the body through passageways of various sizes. He considered that too much constriction causes inflammation, as does peccant material, while too much looseness allows loss of vital material. Proper exercise, food, and drugs maintain or restore the body's economy, which is health.[62] His general viewpoint was, of course, much like that of traditional Greek medicine, but his atomic theory and the dogmata he drew from it gave Asclepiades a standpoint from which he could criticize the theory and practice of all predecessors.

Galen seems to think that Asclepiades' physiology is closest to that of Erasistratus, which seems likely, both because, after Aristotle, Erasistratus was *the* notable predecessor in that field, and also because notions of pressure, flow, blockage and release figure prominently in the theories of both men, while some notions crucial to Galen (such as "attraction" and various categories of "cause") became prominent only later, with the Pneumatics.[63] Asclepiades did write *Refutations* of Erasistratus, but what he said or how he acknowledged his own debts is unknown.[64] Most apparent in Asclepiades' writings is his rejection of predecessors for their murderous ignorance. In his writings, Asclepiades

62. Soranus gives us a brief account of Asclepiades' approach to medicine (Cael. Aurel., *Acut.* 1.14), as well as many examples of his methods of therapy. Christian Gumpert's treatment of the fragments of Asclepiades (Weimar, 1794) has been translated by R. Montraville Green, *Asclepiades, His Life and Writings* (New Haven, 1955). Much of the vitriolic gossip about Asclepiades is preserved by Pliny, *Natural History* 26.6–9.

63. Galen criticizes the physiology of Erasistratus and Asclepiades extensively in *Natural Faculties* and *Usefulness of the Parts*.

64. Cael. Aurel., *Acut.* 2.33, names the book and speaks of its denial of the presence of fever in cardiac disease. Cael. Aurel., *Morb. Chron.* 2.13, suggests that the work was also called *Preparations* and dealt with practical rather than theoretical refutations. We have one report that Asclepiades said that Erasistratus saw what he wanted to see—some membranes growing from the heart that were not there—which Herophilus had not described (Galen K 1.109).

seems to have developed the style of pointing out all his predecessors' errors as a proeme to his own confident solution of the
medical problems to which they had addressed themselves.
(He received his proper reward by being treated in the same manner
by the Methodists, who took over much of his theory but re-
pudiated his dogmatism and criticized him, as he had criticized
predecessors, for his dangerous and harsh remedies.) A reading
of Caelius Aurelianus' translation of Soranus makes it obvious
that Soranus, or his source, has not only taken over Asclepiades'
mode of cataloguing predecessors' inadequacies in definition
and treatment of disease (which we might call *hamartography*),
but has taken over much of the substance of his catalogues as
well, and then appended descriptions of Asclepiades' errors
along with those of Themison and other followers of
Asclepiades.[65] Soranus gives an idea of Asclepiades' manner of
presentation when he describes his treatment of *phrenitis* (Cael.
Aurel., *Acut.* 1.15): first Asclepiades refuted all who prescribed
contrary measures, second he discussed how the disease should
be prevented, and third he indicated how it should be treated if
it occurs. In his own work, Soranus generally alters the order
and puts the refutations last.

Hippocrates was one of the predecessors to be criticized and
replaced by the new dogmatism. But how did Asclepiades con-
ceive Hippocratic medicine? We know specifically that he rejected the Hippocratic notion of particular days which are critical in disease (a notion that the Empirics had embraced and used
for proof of Hippocrates' empiricism).[66] Asclepiades also limited
severely the uses of the standard purges, clysters, and other
preparations prominent in the Corpus. We can infer from
Soranus' hamartography that much of the therapy of Hippo-

---

65. Note, for example, in Cael. Aurel., *Acut.* 2.13, the hamartography of definitions of pleurisy, whose mistakes prepare for Asclepiades' definition in his
work *Definitions*, to which Soranus appends the criticism that Asclepiades'
definition offers the wrong emphasis, and cf. *Acut.* 2.26. Note also that the
report and criticism of Herodicus' (and Euryphon's) treatment of dropsy is
specifically credited to Asclepiades (Cael. Aurel., *Morb. Chron.* 3.8).
66. Celsus 3.4.11. For empiric treatment of the subject of critical days, see Galen, *On Medical Experience*, ch. 21.

crates which Asclepiades criticized appears in *Regimen in Acute Diseases,* including the last part of that work (the so-called appendix or *notha*), which in Galen's time could no longer be confidently said to have been written by Hippocrates (though Galen argues that its substance is Hippocratic).[67] *Prorrhetic, Diseases* 2 and 3, *Affections, Aphorisms,* and other Hippocratic works are drawn on for the hamartography of Soranus.[68] But it would be rash to draw many conclusions about Asclepiades from that list. Celsus criticizes Asclepiades for failing to acknowledge Hippocrates' brief statement of all the principles of massage while acting as though he invented the subject (2.14.2). But in one of the two quotations from Asclepiades preserved by Oribasius (4.2.43-44), Asclepiades says that Hippocrates' words in *Joints* about spontaneous dislocations are true because he has seen the phenomenon himself. Indeed, as I said above, I suspect that Asclepiades' motto, as Celsus cites it (3.4.1), is adapted from *The Surgery,* chapter 7. Asclepiades said that the job of the physician is to cure safely, speedily, and pleasantly (*tuto, celeriter, iucunde*). "Do it quickly, readily, painlessly, neatly," says the Hippocratic text.

We do know that Asclepiades wrote commentaries on two works of the Corpus, *Aphorisms* and *The Surgery.* Of the *Aphorisms* commentary Galen gives no information (Galen's own commentary was written without recourse to predecessors), but Soranus (Cael. Aurel. *Acut.* 3.1) preserves a definition of *synanche* from it. When Galen wrote his commentary on *The Surgery,* he was beginning to do research in other people's commentaries, and he appears to have used Asclepiades' commentary for his information about the early condition of the text and about Hellenistic commentators on it. But Galen cites only three textual readings from Asclepiades and one insignificant interpretation of a passage (Galen, K 18 B.666, 715, 805, 810). Galen's disparaging reference to Asclepiades in his long com-

---

67. Besides Galen's statements on the subject, his contemporary, Athenaeus (*Deipnosophistae* 2.570), said that *Regimen in Acute Diseases* is half or wholly spurious. The fact that the Methodist hamartography unembarrassedly treats the *notha* as genuine is evidence that its basis was laid before the work of Dioscurides, Hippocrates' editor.

68. The index of Drabkin's edition of Cael. Aurel. provides the details.

ment on the work's opening passage may conceal something of interest. The passage in question speaks of sources of knowledge, sight, touch, etc., ending, as Galen reads it, with "all the things by which we know what there is to know." Galen speaks at tedious length of those who have treated the passage as a general theoretical introduction to medicine, an epistemology aimed at disputes between the ancient equivalents of Academics, Stoics, Skeptics, and others. Galen's impetus in his discussion comes from his teacher Aephicianus, who interpreted the Hippocratic passage as the equivalent of Simias the Stoic's theory of knowledge (pp. 654–655). Galen, on the contrary, thinks that the passage foreshadows his own work on the *koinos logos,* which argued that perception, memory, and reason are the three mental faculties (pp. 657–663). In the midst of his exposition, Galen remarks that Asclepiades tried to overturn the criteria of judgment and the reasoning faculty, with many other things, as nonexistent (p. 660). This may be Galen's characteristic lateral abuse, or it may say something of Asclepiades' method of commentary.

In sum, Asclepiades appears not to have read Hippocrates in his own image, but to have begun a tradition continued by the later Methodists: the view that Hippocrates had much to offer, but that his errors were typical of those of the medicine of the benighted past, which the new methodology was to overcome. Not everyone worshiped the medicine of the classical period as Galen did nor tried to read his own views into the past and into "Hippocrates."

After the time of Asclepiades, Rome became the center for creative developments in medicine. In the early Roman Empire the Methodist and pneumatic schools of medicine were developed. Erotian, in the time of Nero (54–68 A.D.), wrote his dictionary for those who would read Hippocratic literature, and in the time of Hadrian (117–138 A.D.), the first scholarly editions of the Corpus were composed by Dioscurides and Artemidorus Capiton, who drastically altered the traditions about the Hippocratic writings, with results for Galen's views which I shall try to assess.

Celsus wrote in the time of Tiberius (14–37 A.D.). His work *On Medicine* (part of an encyclopedic work whose lost portions included agriculture, rhetoric, and military science) gives what appears to me to be the sensible educated man's view of medicine in his time. The source of his material and whether he translated into his excellent Latin from one Greek work or several have been disputed, as has his own experience or lack of it in medical practice.[69] Because the degree of his originality cannot be assessed, my discussion of him must be tentative in various respects. I shall be speaking of views that he accepted or himself compounded from medical writings of his own period. Celsus summarizes current knowledge of disease and therapy from a specific point of view, which he outlines. And he gives his version of the history of medicine and Hippocrates' place in it. His work is important for estimation of the Hippocratic tradition in his time.

Celsus' own view of the history of medicine is quaintly dogmatic in tenor. Whereas the Empirics, as he reports them, envisioned a gradual development of medical practice from observations of phenomena, Celsus believes that before Hippocrates medicine was part of philosophy: philosophers were most likely to get sick because they got little exercise and hence had to think about health and disease. Pythagoras, Empedocles, and Democritus worked at medicine. Hippocrates, Democritus' student, first separated medicine from philosophy. Diocles, Praxagoras, Chrysippus, Herophilus, and Erasistratus advanced the science, which was divided into dietetics, pharmaceutics, and surgery. These practitioners of dietetics claimed knowledge of natural philosophy, but the Empirics who came later denied that such reasoning was pertinent to medicine. Asclepiades next changed methods of therapy, and his successor Themison altered Asclepiades' methods somewhat. Such, in summary, is Celsus' history of medicine before his own time (*Proem.* 7–10).

Celsus' presentation of his own point of view in medicine is

69. See Owsei Temkin's summary of the evidence, and his judicious assessment of it: "Celsus' 'On Medicine' and the Ancient Medical Sects," *Bulletin of the History of Medicine* 3 (1935), 249–264. Celsus' work is available in *Aulus Cornelius Celsus quae supersunt*, ed. Friedrich Marx (Leipzig, 1915), and with translation by W. G. Spencer in the Loeb Classical Library. I cite it according to book, chapter, and section as used in both editions.

cast in the form of a comparison of the views of Dogmatics, Empirics, and Methodists. He elects for himself a modified empiricism: a rational science that treats disease according to its evident causes, as the Empirics do, but which adds to its intellectual perfection by investigating the nature of things, as Hippocrates and Erasistratus did to some extent (*Proem.* 47, 74–75). Methodism he rejects as irrational and inconsistent (*Proem.* 62–67). For the source of the principles of his medicine, Celsus cites Hippocrates: one should take account of both general and particular characteristics in diseases (*Proem.* 66). His lists of possible dogmatic views are interesting: the philosophers say that an excess or deficiency among the four elements causes disease; Herophilus says that the cause is in the humors; Hippocrates, that it is in the pneuma; Erasistratus, that the cause is the entry of blood into the pneumatic vessels; Asclepiades that the cause is blockage of atoms (*Proem.* 14–15). For causes of digestion, he cites the following: Erasistratus: grinding; Plistonicus: putrefaction; Hippocrates: cooking; Asclepiades: none of those (*Proem.* 20–21).

Celsus thus gives a new version of Hippocrates: father of medical science, student of philosophy, and, in effect, source of dogmatic medicine which the Empirics modified. Hippocrates' view of the cause of disease as Celsus gives it seems to go back to Menon's doxography; his tutelage by Democritus goes back to the myths of the pseudepigrapha. The view that medicine was only a part of philosophy before Hippocrates delineated the science is novel (and contradicts Menon), not unrelated to Celsus' statement that those who pursued dietetics also pursued natural philosophy; that is, they were dogmatics.[70] Whence comes Celsus' amalgam? He invented it, perhaps, because collectors of encyclopedic learning needed to show no great historical judgment, or possibly took it from another, conceivably from

----

70. As noted above, the empirics held the "Hippocratic" view of *Ancient Medicine*, that dietetics was a very ancient art based on reasoning from observation and experience. Celsus traces surgery and pharmaceutics back to the time of the Trojan War (*Proem.* 3) and says that surgery was very ancient, but that Hippocrates practiced it more than his predecessors (book 7 *Proem.* 2).

Asclepiades, whose influence shows variously in Celsus' work and whose tendentious dogmatic medical history might have resembled that of Celsus. I have little confidence in such conjectures and prefer to note that Celsus' historical views could be expressed in his time and would have been available after his time. But most significant for my purposes, the views may have been new in the time of Celsus.

Besides presenting a novel Hippocrates and citing him for the principles of his own medicine, Celsus makes extensive use of the Hippocratic Corpus in the body of his work. His presentation of surgical procedures in book 8 is almost wholly a digest from the surgical works of the Corpus. In his presentation of dietetics and in the descriptions of symptoms and treatments of disease, there are frequent unacknowledged quotations from Hippocratic works.[71] The views of Hippocrates, father of the whole medical science (book 7 *Proem.* 2), are sometimes acknowledged explicitly. Celsus chides Asclepiades for failing to credit Hippocrates with saying everything about massage in a few words (2.14.2; Asclepiades wrote as though he had invented the subject. Cf. Galen on Theon, above p. 111). And Celsus holds up Hippocrates as a model of modesty and truth, characteristics of a great man (8.4.3, in reference to a first-person anecdote in *Epidemics* 5). Celsus' medicine appears to be, in accord with his principles, a modified Empiric medicine. He deals with diseases according to syndromes of symptoms, and therapy that has proved effective. He acknowledges and accepts various advances in therapy made by Asclepiades (including his rejection of the "Hippocratic" and Empiric doctrine of critical days, 3.4.11–15), but he also criticizes Asclepiades (and Themison) on Empiric grounds (3.4.7–8). Celsus betrays no sophisticated view about multiple authorship of the Corpus Hippocraticum (notably, he quotes freely from *Prorrhetic* 2), nor knowledge of Hippocrates' sons and students in that or any other connection. He equates Hippocrates with the useful works of the

---

71. A long but incomplete list of passages in Celsus drawn from the Hippocratic Corpus can be found in the index to Marx's edition and a slightly expanded list in Spencer's Loeb Library edition, volume 3.

Corpus, in the manner in which, so I have argued, the Alexandrians did before him. There is as yet no sign of a Hippocratic Question.

The pneumatic school of medicine, which Celsus does not mention, exerted its influence on the Hippocratic tradition indirectly through Galen, who took over much of the system of the Pneumatics and attributed it to Hippocrates. I believe, however, that the early Pneumatics did not attribute their doctrine to Hippocrates nor claim him as their own. Instead, they put themselves in the dogmatic tradition of Alexandria (as had Asclepiades), but they rewrote dogmatic medicine according to the philosophical system of the Stoics. I shall try to outline briefly their approach to medicine and to assess their use of the past.[72]

72. Still the basic book on the pneumatic school is the clear and useful study by Max Wellmann, *Die pneumatische Schule bis auf Archigenes,* Philologische Untersuchungen 14 (Berlin, 1895). The subject needs reworking, partly because Wellmann used spurious material as Galenic, partly because of predilections he brought to the work. Wellmann's article in *RE s.v.* "Athenaios," gives a good digest of Athenaeus' contributions to medical theory. Fridolf Kudlien has contributed to the subject in the following articles: "Posidonius und die Aerzteschule der Pneumatiker," *Hermes* 90 (1962), 419–429; "Untersuchungen zu Aretaios von Kappadokien," *Abhandlungen der Akademie der Wissenschaften und der Literatur,* Mainz, geistes- und sozialwissenschaftliche Klasse, no. 11 (1963), 1151–1230; *RE* Suppl. 11, 1097–1098, *s.v.* "pneumatische Aerzte." I do not accept some of his conclusions about the Pneumatics. Kudlien insists that Athenaeus' relationship to Posidonius was discipleship and that Aretaeus must have written in the first century A.D. The questions remain open, but Kudlien is not convincing because much of his argumentation appears ill-considered. Wellmann's facts remain: Celsus did not know of the pneumatics. Archigenes was creative and influential, Aretaeus was not, and Aretaeus cannot be dated confidently. Kudlien dismisses Wellmann's considerations and uses arguments such as "Aretaeus was not read because he was so rational" (cf. "Untersuchungen," p. 28). The putative Ionic, Hippocratic revival, during which Aretaeus' work and *Nutriment* and *On the Heart* must have been written, Kudlien proves by pointing to the works themselves (cf. *RE* Suppl. 11, 1104), which cannot be dated. Hans Diller, "Eine stoisch-pneumatische Schrift im Corpus Hippocraticum," *Sudhoffs Archiv* 29 (1936), 178–195, showed convincingly that late language in *Nutriment* suggests a date much after the fourth century B.C. His suggestion that it is pneumatic is much less substantial and more tentative, but Kudlien rashly piles more hypotheses on it. Robert Joly's introduction to *Nutriment* in his Budé edition (*Hippocrate,* Tome 6, part 2 [Paris, 1972], 129–138) offers some needed corrections to Diller's and Kudlien's assessments. More needs to be done.

Cosmic unity and sympathy are their first principles. Athenaeus, the school's founder, described a cosmos which is a continuum. Basic matter acted on by hot, cold, wet, and dry produces the world we see. The precise mixture of hot, cold, wet, and dry determines what a thing is, and all things can change into all others with a change of mixture. *Pneuma* in the cosmos and the body holds things together and makes them function. A change in the *pneuma* signals (or causes) a change in the mixture (or temperament) of a thing (animate or inanimate). Health is a proper mixture or temper that accords with the Nature of each thing, *eucrasia. Dyscrasia,* departure from nature, brings illness, dissolution, and death. This new synthesis of many old ideas gave the Pneumatics a lens through which they could reexamine traditional medicine and a vocabulary with which they could rewrite it.

Their individual contributions are sometimes difficult to distinguish. Athenaeus certainly produced the manifesto and worked out the cosmology. His large work on *Healing (Boethemata,* at least thirty books) dealt with temperaments of persons, seasons, places, foods, ages of life, and probably drugs and was an attempt to analyze the world and man for medical purposes. I cannot guess from the remaining fragments how detailed his physiology and anatomy were. Archigenes, whose therapeutics was influential, made elaborate classifications of pulses and fevers for diagnostic purposes. Others worked on surgery. Agathinus, teacher of Archigenes, had expanded Athenaeus' doctrines, or perhaps the scope of his medicine, by including material from Methodists and Empirics. He was called an Eclectic or Episynthetic. Leonidas the surgeon was called an Episynthetic. How members of this centrifugal school talked about themselves is not entirely clear from the meager remains of their work, but some notion of their use of the past can be gotten from Galen's reports and from the preserved fragments.

Athenaeus declared himself a dogmatic, but specifically rejected Asclepiades' cosmology and his reasoning (Galen, K 1.486, cf. 14.676). For their notions of temperaments and seasons, as for their elemental theory, Athenaeus and other pneumatics adduced the authority of Aristotle, Theophrastus, and the Stoics (Galen, K 1.522–35). We do have the following quotation from Athenaeus: "The elements of medicine are, as some of the ancients supposed, the hot, the cold, the wet, and the dry, from which first, simplest, and least phenomena man is composed, and into which finally he is dissolved" (Galen, K 19.356). When Athenaeus discussed trembling, palpitation, etc., the only predecessors he cited (in order to refute them) were Asclepiades and two philosophers, an Academic and a Peripatetic (K 7.614, cf. 7.165–166). Archigenes' scheme of diagnosis by affected places (which Galen took over) was a novelty in which Archigenes seems to have acknowledged no predecessors (K 8.136), except perhaps to correct Erasistratus on the subject of semeiosis and crises in disease (K 9.668–669). The elaborate pulse lore of Archigenes was a correction and extension of what Herophilus began, but the Pneumatic Magnus was the immediate source whom he acknowledged and corrected (K 8.638, 9.8, 18; cf. Wellmann, *Die Pneumatische Schule*, pp. 172–201). Athenaeus' embryology was a direct extension of that of Aristotle (K 4.593 f., cf. Wellmann, pp. 100–104). Pneumatics' logical refinements of notions of cause are Stoic corrections of the Alexandrians (cf. CMG suppl. 2, Intro.).

In short, for their dogmatic, philosophical medicine they claimed originality or sought their antecedents in Alexandrian dogmatism and in philosophy. In that they were like Asclepiades. A reading of the preserved fragments (mostly from Oribasius' medical collection) confirms my impression that the Pneumatic theorists virtually ignored Hippocrates and made little direct use of the Corpus. Athenaeus, writing on healthy and unhealthy places and waters, for example, deals with the same material that is in *Airs Waters Places,* but does so in his own way and from his own viewpoint without allusion to the Hippocratic treatise (Orib. 2.291–306). So it is, generally, with Athenaeus, Archigenes, and even Antyllus. Athenaeus quotes Hippocrates once in his hygiene, calling him "the ancient (*palaios*) Hippocrates": "Thought is exercise for the soul" (Orib. 3.98 from *Epidemics* 5.31). Similarly, he cites Empedocles for the mathematics of gestation (Orib. 3.78; one notes that he does not cite Hippocrates on this subject) and quotes the physicians Diocles and Andreas (Orib. 3.78, 108).[73]

I point out and emphasize the Pneumatic theorists' lack of attention to Hippocrates in order to correct past habits of reading medical history through Galen's eyes. It is not true that, as Galen saw it, everyone was a "follower" or "enemy" of Hippocrates and so oriented his medicine. The Pneumatics must have been aware of books of the Corpus, but they do not appear to have claimed that those books contained their science of elemental *eucrasia* and *dyscrasia*.[74] The tradition was much more various than Galen would have it and than scholars have realized. The evidence suggests that only Galen and his immediate Hippocratean predecessors reconciled pneumatic and "Hippocratean" medicine.

Contemporary with the work of the early Pneumatics, we have evidence of a burgeoning Roman tradition of literary study of the Hippocratic writings which was part of a general literary revival in first-century Rome. The period cannot glibly be called an age of archaism—first because archaism was endemic in antiquity, second because literary study of Hippocrates and Celsus' use of him are contemporary with Thessalus' assertion of revolutionary Methodism, as well as with the Pneumatics' new system of medicine.[75] Nor, so far as we can fairly infer, did

73. Wellmann, *Die pneumatische Schule,* page 10 and note 3, speaks of Athenaeus' doxographic interests, but that is a severe overstatement of the case. Galen (K 7.614) shows irritation at his lack of doxographic interest.

74. The remains of the Pneumatics show little concern with the humors, but much with the four qualities. Galen's work, when he draws on them, shows the same emphasis (for example, in *Differentiation of Diseases,* K 6.836–880). I suspect that strictly humoral pathology had little place in their system and that its prominence in later medicine came from "Hippocratization" of their theories. The sensible book by Erich Schoener, *Das Viererschema in der antiker Humoralpathologie, Sudhoffs Archiv,* Beiheft 4 (Wiesbaden, 1964), makes progress toward reconstruction of the historical sequence that led to the humoral pathology of late antiquity.

75. For Galen's irritation that Thessalus claimed to be doing something new in the world, see K 10.7–8. Indication that Pneumatics competed with Methodists in novelty is the book by the Pneumatic Magnus, which had the intriguing title *Discoveries since Themison's Time* (K 8.640).

Hippocrates represent all of medicine, or the best in it, for Erotian and the editors of the Corpus.

Erotian composed a much larger Hippocratic dictionary than the excerpts that remain. Scholarly labor, culminating in the work of Ernst Nachmanson, has clarified the outlines of the original work, which explained words from approximately thirty-seven treatises in the sequence in which the words appear in the treatises, probably following the form of Bacchius' work three centuries earlier.[76] We do have Erotian's preface, fortunately, in which he reviews his sources and sets forth his principles. His purpose is to make Hippocrates available to his age, and he recommends him as good literature, particularly for medical men who can test their knowledge against his writings. Modest and charming about his author, Erotian still insists that his dictionary will be the best one yet done. Erotian's preface also lists (with some omissions) the works he will gloss, and he adds that *Prorrhetic 2* is not by Hippocrates, as he will show elsewhere (he did not treat it in his dictionary). Considering the breadth and catholicity of his list, from pseudepigrapha through *Epidemics, Aphorisms, Prognostic,* works on disease, gynecological works, to *Nutriment* and *Ancient Medicine,* it seems odd that he singles out *Prorrhetic 2* as not Hippocratic.[77] I know of no explanation for Erotian's statement, and scholars until now have been unaware of what I have pointed out above: this is the first known statement in antiquity that a particular work of the Corpus was not by Hippocrates. There are no apparent linguistic or stylistic grounds, nor any doctrinal reason. This is the only hint that Erotian may have held a particular view of Hippocrates beyond thinking him a venerable, classical medical writer, who wrote in interesting archaic language and dialect. It may be that his predecessors had neglected to gloss *Prorrhetic 2.* Leaving that problem unsolved, we may pass on to Dioscurides and Capiton, editors of Hippocrates, who made many judgments about genuineness and who exerted great influence on the tradition.

76. Ernst Nachmanson, *Erotianstudien* (Uppsala, 1917), and *Erotiani vocum Hippocraticarum collectio.*
77. It is possible, as Littré says (vol. 1, p. 410), that Erotian's words refer to *Prorrhetic 1* as well, but unlikely considering Nachmanson's results.

Around the beginning of the second century A.D., Dioscurides and Artemidorus Capiton produced scholarly editions of the Hippocratic works in obvious imitation of the Alexandrian editors of classical Greek literary texts, and in working on Hippocrates they worked on an author whom their Alexandrian predecessors had passed over.[78] Dioscurides, who seems to be the earlier of the two, imitated Aristarchus in judging the mind and style of his author and in using an obelus mark in the margin to indicate passages he thought to be spurious.[79] Like the Alexandrians, Dioscurides was concerned about his author's decorousness, and he obelized a passage in *Epidemics* 6 on the grounds that Hippocrates would never deceive a patient. (Hence, he judged, Hippocrates' son Thessalus must have written it [CMG 5.10.2.2, p. 283]). Dioscurides and Capiton after him altered the language of Hippocratic treatises to the dialect of Cos, as they understood it.[80] Some material which Dioscurides thought was intrusive he conjectured must originally have been a marginal comment which the copyist ignorantly entered in the text (CMG 5.10.2.2, p. 464). Some passages that he condemned, Dioscurides (and Capiton after him) omitted from his edition; some he relegated to the upper or lower margin (CMG 5.9.2, pp. 176, 243). Galen sometimes speaks of the two men together, sometimes of one, and sometimes indicates that they handled

---

78. The most comprehensive study of their editions is that by Johannes Ilberg, "Die Hippokratesausgaben des Artemidorus Kapiton und Dioskorides," *Rheinisches Museum* 45 (1890), 111–137. Ilberg directs attention mainly to the textual tradition. Some corrections were made by Franz Pfaff, "Die Ueberlieferung des Corpus Hippocraticum in der nachalexandrinischen Zeit," *Wiener Studien* 50 (1932), 67–82.

79. It is Galen who specifies that Dioscurides imitated Aristarchus: CMG 5.10.2.2, p. 283 (on *Epidemics* 6; cf. on *Nature of Man* CMG 5.9.1, p. 58). Virtually all our information about Dioscurides and Capiton comes from Galen's Hippocratic commentaries, to which my references in the text pertain unless I specify otherwise.

80. CMG 5.10.2.2, pp. 6 and 483. Elsewhere Galen speaks of "some" who said that Hippocrates' dialect was old Attic (K 18 B.322.10, on *Fractures*). Sometimes he simply speaks of the dialect as Ionic, as in CMG 5.9.1, pp. 193, 236, 260. Galen seems not to think highly of their use of the "Coan" dialect and at times points out their ignorance of linguistic usage (CMG 5.10.1, pp. 175, 197).

passages differently.[81] Galen shows his irritation at Dioscurides
and Capiton variously: frequently he condemns their "reckless"
alterations of the text, which alter meanings he would like to
preserve; once, in irritation at their superficial handling of a
superfluous particle, *men,* Galen remarks that they pretend to be
better grammarians than anyone else (CMG 5.10.2.2, p. 83). I
suspect, though without proof, that Dioscurides' edition was re-
done by Capiton (his relative, relationship unspecified [CMG
5.9.1, p. 13]), that Capiton's edition indicated the contributions
of each, and that it was Capiton's edition from which Galen got
his information about both.[82]

Dioscurides and Capiton were the first to produce scholarly
literary editions of Hippocrates. The significance of that fact
and the nature and influence of their work have not been much
considered by modern scholars, who have generally emphasized
textual emendations and accepted unfounded generalizations
about Alexandrian *Echtheitskritik.*[83] My study of the tradition to
this point makes it possible for me to offer some rather startling
observations. Dioscurides (if he was the earlier of the two)
undertook a task no one before him had done: systematically to
make sense for the sensible litterateur of the disparate material
of the Corpus. He had the glossographers and grammarians as
predecessors (whence his feeling for Coan dialect), and he had
the pseudepigrapha to tell him of his author's character, life,

81. For the readings of each which Galen reports, see Ilberg, "Die
Hippokratesausgaben," and Pfaff, "Die Ueberlieferung." For distinctive handling of a passage in their two editions, see CMG 5.9.2, p. 131.
82. Ancient literary economy, partly caused by the unwieldy rolls of papyrus,
seems to require that Galen used a combined edition which distinguished
the two scholars' judgments. Galen tends to speak of them as Artemidorus
and Dioscurides in his earlier commentaries and as Dioscurides and Artemidorus or Dioscurides and Capiton in his later commentaries; I do not
know why. With great consistency, the medieval manuscript tradition of the
Corpus has followed the readings of Capiton where Galen reports them.
83. Kurt Bardong's interpretation of the phrase "from the little *pinax*" in
Epidemics 6 gives a good example of the direction scholarly reasoning has
taken. He also reports much of the bibliography on the subject: "Beiträge
zur Hippokrates- und Galenforschung," *Nachrichten von der Akademie der
Wissenschaften in Göttingen,* philosophisch-historische Klasse, no. 7 (1942),
577–640.

and family: eloquent and rational Hippocrates went to school with Democritus and Gorgias, came from a line of Asclepiads, and left sons, grandsons, and students. Dioscurides also had such Alexandrian precedents as conjectures about the Homeric corpus and conjectures that some plays of Euripides and some of Sophocles were partly reworked and partly rewritten by their sons and literary executors. We can get a fair idea how he proceeded to his task of editing and interpreting his author's text.

Dioscurides attributed *Diseases* 2 to another Hippocrates, the grandson of the "great" Hippocrates; he did the same with part of *Nature of Man* (CMG 5.10.2.2, p. 55, *CMG* 5.9.1, p. 58). He thought that the indecorous passage of *Epidemics* 6 must have been added by Thessalus, the son and literary executor (CMG 5.10.2.2, p. 283). Beyond that, Galen's specific attributions of opinion give out, but Galen alludes variously to a consistent series of stylistic judgments and attributions of works which I think must go back to the work of Dioscurides or "those about Dioscurides." Dioscurides is *absolutely the earliest* person who is reported to have made such judgments. Besides that, he is Hippocrates' first literary editor. The judgment that Thessalus wrote parts of *Epidemics* 6 depends on the stylistic characterization of the *Epidemics* as a whole, which Galen attributes to "those who know best about these things" (K 7.890). "Their" view is that Thessalus put *Epidemics* 2 and 6 together out of notes left by his father on scraps of paper, skin, and wax tablets, whence the different style of those books in comparison with *Epidemics* 1 and 3, which were finished by Hippocrates for publication (CMG 5.10.2.2, p. 76; CMG 5.10.1, pp. 213, 310; K 7.854–855, etc.). *Epidemics* 5 was attributed to Hippocrates the son of Draco, the grandson of the great Hippocrates (K 7.854). "People best able to estimate the force of Hippocrates' books" considered *Epidemics* 4 to be part of what Thessalus put together, but Galen disagreed (K 7.891; CMG 5.10.2.2, p. 76). Dioscurides and Capiton, I suspect, are the people whom, in another mood, Galen resents because they "rob" Hippocrates of works that are precious to him. They jealously robbed Hippocrates of the last part of *Regimen in Acute Diseases* and of other works, but did not dare to rob him of the earlier portion (K 7.913). We have observed above that the hamartography that stemmed from Asclepiades unembarrassedly treated the whole of *Regimen in Acute Diseases* as Hippocratic, as did Celsus. *Prorrhetic* 1, which Galen was concerned to "save" early in his career, but wished later to damn as spurious, appears to have been judged "clearly of Hippocrates' craft, but inferior," therefore by one or the other grandsons of the great Hippocrates.[84] Hippocrates, the grandfather of the great Hippocrates, was suggested as author of the large surgical work that comprised both *Fractures* and *Joints* (CMG 5.9.1, p. 135). This much inference about their judgments seems quite safe.

We can go yet one step further, I think, and attribute to Dioscurides the canon of "unquestioned," or "most genuine and most useful works" (CMG 5.10.2.1, p. 60), the surgical works, and *Aphorisms, Prognostic, Regimen in Acute Diseases, Airs Waters Places,* and *Epidemics* 1 and 3 (this may or may not be the whole list. Galen speaks of it only allusively.).[85] Since Dioscurides assigned the "aberrant" works of the Corpus, he must have had a canon, and there seems no question that this was it. How did he arrive at it? It seems apparent, considering the tradition as I have laid it out, that Dioscurides chose his own canon according to the works which the Empirics had used and commented on—the works which were "most useful" for the Empirics, who began the tradition, became "most genuine and most useful" for Dioscurides and Galen.[86] The equation is not complete, since the Empirics used all the *Epidemics* and also freely used the *notha* of *Regimen in Acute Diseases.* Dioscurides pared down the Empiric list on stylistic grounds. But, in going to the roots of the tradi-

84. CMG 5.9.2, p. 68. Galen's early affection for *Prorrhetic* may come from the fact that his teacher Satyrus had accepted it as genuinely Hippocratic (CMG 5.9.2, p. 20, and cf. Littré, vol. 5, p. 512).
85. We can see the logic behind the attribution of surgical works to the time before Hippocrates in such histories of medicine as that of Celsus, according to which surgery and pharmacy were very old, but rational medical science was invented by Hippocrates (*Proem.* 1–8). But Celsus spoke of the surgical works as Hippocratic. Dioscurides took the next step.
86. "Usefulness" of texts and of his own commentary on them was the criterion that guided Heraclides of Tarentum. Hence Galen was surprised that he wasted his time on the symbols in *Epidemics* 3 (CMG 5.10.2.1, p. 87).

tion, Dioscurides was, in essence, going back to the Empirics. Hence the "most genuine" works of the Corpus are precisely those which are full of observations of phenomena of disease and cure, the least theoretical ones.[87] That fact stands out in Galen's reports, even while Galen is trying to impose theoretical, dogmatic views on Hippocrates.

I wish that I could give a confident description of the physical appearance of their two editions or that of Capiton. Certainly they were on many rolls of papyrus, with marginal indications of editors' views. Dioscurides and Capiton did not write commentary (CMG 5.10.2.2, p. 407), but did they write brief *hypotheses* or introductions to precede the works, perhaps with their own opinions? Did *Regimen*, Hippocrates' own work, appear with something like the following?

> *Regimen, by Hippocrates. Or perhaps by Philistion, Ariston, Euryphon, Philetas, or another. This work is clearly ancient. When this book is transmitted as a whole in three parts it is entitled the *Nature of Man and Regimen*. When the latter part is transmitted separately it is called *Regimen* or *Regimen in Health*. [Cf. Galen, K 6.473, CMG 5.9.1, p. 135, and p. 59 above).

I will not speculate further here. I offer this putative example to explain the possible source of Galen's ponderous, uninformative (and in part uninformed) statements about "some say... and others say...." I think that Galen's statements must depend on some such *hypotheses*.[88]

I have proposed what many will find startling: that Dioscurides, followed by Capiton, not only assembled the Corpus as

---

87. The Empirics interpreted *Regimen in Acute Diseases* by adducing *Airs Waters Places* (CMG 5.9.1, p. 68). They argued that *Aphorisms* was empiric and adduced *Airs Waters Places* and *Epidemics* as illustration (see Deichgräber, *Empirikerschule*, frags. 362, 363). For the Empirics' use of the Corpus, see above, p. 208.

88. As I have indicated above, p. 161, Dioscurides also wrote a Hippocratic Glossary (probably with an introduction), which Galen discusses in the introduction to his own. Dioscurides showed an excessive (to Galen) interest in animals, fish, places, star lore, etc., but he probably covered all that Galen's work includes (K 19.63–64).

an entity (all agree to that) but that Dioscurides first launched the *Echtheitskritik*, on stylistic grounds and using the mythical genealogies of the pseudepigrapha. My evidence is Galen's ignorance of any previous judgments of the kind and the absence of such evidence from elsewhere: Erotian, who should have shown evidence of previous traditions of that sort, did not (though he thinks one work spurious), nor does anyone else. The next physician and scholar whom we consider will offer some confirmation. Rufus of Ephesus, gentlemanly physician with a bent for antiquarian scholarship in medicine, shows no evidence of concern about *Echtheitskritik* or of taste for controversy. He was roughly contemporary with Dioscurides, perhaps slightly earlier.

Rufus probably practiced and wrote somewhere in the eastern Mediterranean.[89] A fair portion of his extensive and influential writings is preserved. His Hippocratic commentaries are lost, save for some few notices of them by Galen, who recommended him (*Scr. Min.* 2.87). Rufus appears from his language to be a dogmatic: he deals in causes and in humoral pathology and allopathy by hot, cold, wet, and dry. But he does not speak about himself and his views, always directing attention to substantive subject matter. He shows no desire to quarrel with anyone, though he offers corrections to opinions with which he disagrees.[90] His writings suggest such adjectives as gentle, decent, helpful, competent. He seems to write with an eye on the Hip-

89. The remains of Rufus' writings (incomplete) were presented in the useful edition by Charles Daremberg and Emile Ruelle, cited above in note 26. Daremberg unfortunately did not live to make his final contributions to it. Max Wellmann offered a useful characterization of Rufus' medicine (along with conjectures about his sources and him as a source) in "Zur Geschichte der Medicin im Altertum," *Hermes* 47 (1912), 4–17. Material recently recovered from the Arabic has made a new study of Rufus desirable. For Rufus' date and place of practice, see Wellmann, pp. 4–6. Rufus' *Medical Interview* is available with a commentary and German translation in the Corpus Medicorum Graecorum, Suppl. 4, ed. Hans Gärtner (1962), and the text is available in the Teubner series by the same editor (1970).

90. When Galen observes that Rufus was the best of recent writers on black bile, he adds that Rufus is not quarrelsome like Erasistratus, Asclepiades, and the Methodists (CMG 5.4.1.1, p. 71).

pocratic Corpus, not hesitating to disagree with it, but always expressing reverence. For example,

> I hope to offer an important contribution to medicine by showing in the present account which conditions one would restrain with harm when they come on, and which ones one should encourage with irritation rather than stop them. People will say that these are not my discoveries, since Hippocrates long ago often said many such things. And I agree. What is there not in his writings? Still, to put it all in one account, including what has been discovered later, and to treat each thing precisely will make for a not unwelcome treatise. [Orib. 4.83]

Similarly, at the end of his treatise the *Medical Interview,* which tells, step by step, how to elicit from a patient (or his relatives) information necessary to differential diagnosis, he adds this final defense, which may suggest that he was responding to some sort of current tyranny in Hippocrates' name.

> If anyone says that I disagree with Hippocrates, who claimed to have discovered a technique whereby a physician arriving at a city he does not know can learn about the waters and the seasons, the conditions of the digestive systems of the populace, whether they are great eaters or drinkers, and about the common epidemic diseases, how the women are for childbearing, and all else that he attributes to the science, without asking any of the inhabitants, but by his own observation: if anyone brings this up and criticizes me for disagreeing with the physician who was the greatest in the most important matters, then I tell him that I depreciate none of what Hippocrates said, but that some things are discovered also in this way about the climate, the body type, kind of diet, goodness and badness of public waters, and general types of diseases, especially those that are peculiar and strange in each place. I admire the man's wisdom and his excellent discoveries, but I urge anyone who is going to get full and accurate knowledge not to neglect interrogation. [CMG Suppl. 4, pp. 72-73]

Had Rufus dealt with some arrogant Hippocrateans in his day? Did they claim it a virtue to practice just as Hippocrates did? It

242 The Hippocratic Tradition

seems likely. In his general treatise on the symptoms and treatment of diseases of the kidney and bladder, Rufus says:

> Euryades of Sicily and Hippocrates thought that one should operate on patients with nephritis. They advised one to incise the kidney which has a stone or is suppurated, and so to heal it. I cannot disbelieve him [Hippocrates], who is otherwise so competent in the science, but I myself say that I have never dared such a thing. In phthisis, cutting into the thorax by the lower ribs and penetrating within often does work well, by draining the inner lesion. It may be that in desperate cases one should try something of the sort. But if other treatments are effective one should avoid extreme remedies.[91]

Rufus alludes, at the end of the quote, to *Aphorisms* 1.6: "In extreme cases, extreme remedies are most appropriate." I will not pursue the theme of Rufus' struggle with the tyranny of Hippocratism, because, as I said, he does not explain himself except by such delicate and indirect jousting, nor does he characterize the world to which he addresses himself. Still, I cannot help seeing, behind his apologies, implicit criticisms of aggressive, know-nothing Hippocrateans. I am not thinking of Galen, who was aggressive enough in Hippocrates' name, but of the Alexandrian Hippocrateans whom Galen talks about, who read and purveyed Hippocrates without knowledge or subtlety, and whom Galen would like to see whipped for it (cf. above, p. 72).

Whatever his view of Hippocrateans, Rufus appears sincerely affectionate toward Hippocrates. His own account of healthy waters, for example, begins with a series of allusions to and imitations of *Airs Waters Places* and thence proceeds to more elaborate and modern considerations (Dar.-Ru. pp. 341–348). One may ask what Rufus means by "Hippocrates," aside from venerable "father of medicine" (Orib. 2.137), and (as we have seen above) "most excellent physician in most important matters," whose works contain virtually all medicine. Aside from

---

91. Rufus, pp. 20–21, Dar.-Ru. Ruelle translates the reference to Hippocrates somewhat differently than I do.

Internal Affections, to which he alludes on the subject of cutting the kidney, Rufus cites Hippocrates' terminology from surgical works (he seems to have written on the Hippocratic bench, Dar.-Ru. p. 305), from gynecological works, and from *Epidemics* 2, in his work on *Medical Terminology* (Dar.-Ru. pp. 143, 144, 148, 155, 160). He wrote commentary on some of the *Epidemics* and on *Prorrhetic* 1, at least. I would assume that he did more, but I think that we cannot be confident about it.[92]

Galen quotes Rufus' commentaries only a few times, but with respect for his judiciousness.[93] It is clear from Galen's quotations that Rufus was generous with information about past commentators, textual variants, and interpretations of what Hippocrates probably meant.[94] I have referred to Wellmann's (probably erroneous) hypothesis that Rufus, through excerpts from him in Sabinus' commentaries, was the source of all of Galen's information about such matters (above, p. 153). I have noted above, also, that Rufus' information goes back to the Empiric commentators and stops with Bacchius' glossography (above, p. 154); that is, the depth of his tradition is what I have outlined above.

My survey from the fourth century B.C. to the time of Galen's teachers concludes with Rufus, save for one physician, who cannot be neglected but is difficult to place, Aretaeus of Cappadocia, whose work on acute and chronic diseases has been preserved. Where he worked or when is uncertain.[95] He is a very congenial author, whose detailed descriptions of diseases have been much appreciated (by Boerhaave, among others). He was

---

92. For putative commentaries on *Aphorisms* and *Airs Waters Places*, see Dar.-Ru., p. xxxv, Wellmann, "Zur Geschichte der Medicin," p. 9. Evidence for a commentary on *Humors* comes from the forged Galenic commentary on that work.

93. Quotations that show the flavor of Rufus' commentaries can be found in CMG 5.9.2, p. 73, CMG 5.10.2.2, p. 411, and a paraphrase in CMG 5.10.2.2, p. 174.

94. A similar procedure is apparent in Rufus' discussion of icterus, in which he explains why Hippocrates would associate it with critical days (Dar.-Ru., pp. 377–379).

95. See note 72, above, for references to Wellmann's and Kudlien's treatments of Aretaeus. I incline to Wellmann's dating, making Aretaeus roughly Galen's contemporary. Aretaeus is cited according to the edition of Carolus Hude (CMG 2, *editio altera* [Berlin, 1958]), by page and line.

obviously influenced by pneumatic medical theory. Like Rufus, he deals with substantive matters in his works and does not discuss himself, but he is even more reticent than Rufus and offers no polemics or apologies. He writes in Ionic dialect, definitely a literary affection because Ionic had not been spoken for centuries. One presumes that he is imitating the Ionic medical writings, that is, the Corpus Hippocraticum. He cites no authors except Hippocrates, once, and Homer, once.[96] But in the course of his work he quotes from both Homer and Hippocrates a number of times without acknowledgment, simply integrating their words with his text.[97]

From Hippocrates, Aretaeus quotes such sentiments as "not everyone can be cured" (158.6, from *Prognostic* 1), but also such sentiments as "convulsions following a wound are a fatal sign" (5.26, 108.28, from *Aphorisms* 5.2). To the latter Aretaeus adds the humane sentiment, "but one should do what one can" (108.28). Aretaeus contradicts some Hippocratic prognoses to which he alludes (39.10–11, 57.5). He uses a quotation from *Nutriment* that is important to Galen: "The rooting of the veins is the liver" (27.6, 136.10), but he uses it casually, without the implications for physiological theory with which Galen weighted it. Aretaeus quotes or alludes to a wide selection of works from the Corpus. His literary affectations acknowledge The Poet and The Physician, and their dialect, yet his work is not simply a literary exercise but a serious, reasoned medical work, comparable, I suppose, to some hexameter verses Rufus wrote describing laudanum (Galen, K 12.425) or to Democritus' (first century A.D.)

96. Aretaeus cites Hippocrates (p. 44.14) for his use of the word *apoplektos*, in the vicinity of an unacknowledged quotation of *Aphorisms* 2.42: "A violent attack of apoplexy is impossible to cure, a slight one is not easy." He cites Homer to show the venerable association of bile with anger (p. 39.21).

97. In his original CMG edition Hude noted *similia* to various medical authors. In the reissue of the text, Kudlien has added *testimonia et similia*, primarily references to Homer and Hippocrates. The lists are useful, but one wants serious discussion of Kudlien's criteria for relevance. If Aretaeus (15.4–7) says that if the breath is stopped, one dies, and the Hippocratic treatise *Breaths* says it in very different language, what is implied by Kudlien's citation of the passages (173.22)? Too much of Kudlien's list is like that, unfortunately, and does not reflect the true extent of Aretaeus' use of the Corpus.

iambic verses on antidotes (Galen, K 14.117), or to other literary exercises in scientific subjects from the Hellenistic period onward. Aretaeus thus shows an aspect of the Hippocratic tradition. He does not seem to have been influential until long after his own time. Aretaeus' point of view may well have been like that of the Stoic-influenced Aephicianus, Galen's teacher, but Aretaeus does not explain himself.

This survey has brought us back to the time of Galen's teachers, "the time of my father and grandfather," as Galen used to say. I shall conclude with a resumé of where Galen's immediate predecessors fit into the Hippocratic tradition that I have sketched.

To Rufus, Hippocrates is the great father of medicine, whose works contain almost everything. Rufus' audience probably thought it sinful to criticize Hippocrates or to fail to honor him. Perhaps the Eastern tradition of Hippocratizing was stronger than that in Rome. Perhaps in Alexandria a continuous tradition from the time of Apollonius of Citium had claimed to practice medicine just as Hippocrates did, while its practitioners kept up to date by reading into Hippocrates the medical advances that occurred. To posit such a tradition would be pure speculation. I would have to go further and say that the people involved are unknown because they were antihistorical: each communed with the Physician himself, as Galen did, and they left no significant commentary and no history of themselves. (We do not know what was in Areios of Tarsus' *On the Hippocratic Sect,* first century A.D., except for a genealogy [CMG 4.175].). If there was an Eastern tradition of Hippocrateans, they were ineffectual, except for their possible influence on Celsus and Aretaeus, and, of course, Rufus and Galen's teachers.

Hippocrateans there were in the time of Galen's father and grandfather. Sabinus, "clearer than Aristotle" in his explanations, but "without a dream of anatomy," found in Hippocrates a teleologist and humoral therapist. Stratonicus, one of Sabinus' students, was the best clinical therapist of Galen's teachers, while another of Sabinus' students, Metrodorus, rouses Galen's bile and reminds him of the naive Alexandrian Hippocrateans. Sabinus took over the conjectures of Dioscurides about the authors of the works of the Corpus. He seems to have integrated at least some pneumatic medicine with his own humoral therapy and attributed the results to Hippocrates. Did Sabinus' commentaries embody Rufus' results, so that Galen did not have to use Rufus directly? That theory seems untestable: Sabinus is never quoted as even mentioning Rufus.

There were many private views about Hippocratic science among Galen's teachers and their teachers, the esoteric, unpublished, as opposed to exoteric, published ones. Quintus' views were wholly esoteric. Quintus was probably a great man, and he had wit: he was quoted as saying that the virtue of massage with oil was that it removed the clothing and also that "hot and cold" were the names of baths (CMG 5.4.2, p. 100). The Hippocratic interpretations offered by Quintus and by Marinus and Numesianus appear to have been restrained compared to those of Sabinus: they did not find everything in Hippocrates, including patients' addresses, laden with significance; they did not attribute a full panoply of modern dogmatic theories to Hippocrates. They did insist on the phenomena, particularly the phenomena of anatomy, as the basis of medical thought, and although Galen embraced and advanced their anatomical studies, he had to reject their restraint in reading Hippocrates as too empirical. But Aephicianus, Quintus' student, who leaned to Stoicism, read Stoic and Pneumatic lore into Hippocratic texts. Galen corrected him and Sabinus as well by taking the warrant to read his own Platonizing version of Stoicism into Hippocrates.

BIBLIOGRAPHY

I. Texts of Ancient Authors

ANONYMUS LONDINENSIS. *Anonymi Londinensis ex Aristotelis Iatricis et aliis Medicis Eclogae.* Ed. Hermann Diels. *Supplementum Aristotelicum* 3.1. Berlin, 1893.
———. *The Medical Writings of Anonymus Londinensis.* Trans. William H. S. Jones. Cambridge, 1947.
APOLLONIUS OF CITIUM. *Commentarius in Hippocratem De Articulis libri III.* Ed. Jutta Kollesch and Fridolf Kudlien. Trans. Jutta Kollesch and Diethard Nickel. Corpus Medicorum Graecorum 11.1.1, Berlin, 1965.
ARETAEUS. *Aretaeus.* Ed. Karl Hude. CMG 2, 2d ed. Berlin, 1958.
CAELIUS AURELIANUS. *On Acute Diseases and On Chronic Diseases.* Ed. and trans. Israel E. Drabkin. Chicago, 1950.
CELSUS. *Auli Cornelii Celsi quae supersunt.* Ed. Friedrich Marx. Leipzig, 1915.
———. *De Medicina.* Ed. and trans. W. G. Spencer. 3 vols. Loeb Classical Library. London, 1935–1938.
EROTIAN. *Erotiani Vocum Hippocraticum collectio.* Ed. Ernst Nachmanson. Uppsala, 1918.
GALEN. Corpus Medicorum Graecorum. Academiae Berolinensis Hauniensis Lipsiensis.
 5.2.1. *De uteri dissectione.* Ed. Diethard Nickel. 1971.
 5.4.1.1. *De propriorum animi cuiuslibet affectuum dignotione et curatione. De atra bile.* Ed. Willem de Boer. 1937.
 5.4.2. *De sanitate tuenda libri VI.* Ed. Conrad Koch. *De alimentorum facultatibus libri III.* Ed. Georg Helmreich. *De bonis malisque succis.* Ed. Georg Helmreich. *De victu attenuante.* Ed. Karl Kalbfleisch. *De ptisana.* Ed. Otto Hartlich. 1923.

5.9.1. *In Hippocratis de natura hominis commentaria III.* Ed. Ioannes Mewaldt. *In Hippocratis de victu acutorum comment. IV.* Ed. Goerg Helmreich. *De diaeta Hippocratis in morbis acutis.* Ed. Ioannes Westenberger. 1914.

5.9.2. *In Hippocratis Prorrheticum I commentaria III.* Ed. Hermann Diels. *De comate secundum Hippocratem.* Ed. Ioannes Mewaldt. *In Hippocratis prognosticum commentaria III.* Ed. Iosephus Heeg. 1915.

5.10.1. *In Hippocratis epidemiarum librum I commentaria III.* Ed. Ernst Wenkebach. *In Hippocratis epidemiarum librum II comment. V ex versione Arabica.* Ed. and trans. Franz Pfaff. 1934.

5.10.2.1. *In Hippocratis epidemiarum librum III commentaria III.* Ed. Ernst Wenkebach. 1936.

5.10.2.2. *In Hippocratis epidemiarum librum VI, Commentaria I-VI.* Ed. Ernst Wenkebach. *Comment. VI-VIII ex versione Arabica.* Ed. and trans. Franz Pfaff. 1940.

5.10.2.3. *In Hippocratis epidemiarum comment. Indices nominum et verborum Graecorum.* Prep. by Ernst Wenkebach and Konrad Schubring. 1955.

5.10.2.4. *Kommentare zu den Epidemien des Hippokrates.* Indices der aus der Arabischen übersetzten Namen and Wörter. Prep. by Franz Pfaff, Karl Deichgräber, and Fridolf Kudlien. 1960.

5.10.3. *Adversus Lycum et adversus Julianum libelli.* Ed. Ernst Wenkebach. 1951.

Suppl. 2. *Galeni De causis procatarcticis libellus a Nicolao Regino in sermonum Latinum transl.* Ed. Kurt Bardong. 1937.

Suppl. Or. 2. *Galeni De diaeta in morbis acutis. De causis continentibus. De partibus artis medicativae.* Ed. and trans. from the Arabic by Malcolm Lyons. 1969.

――. *Claudii Galeni Opera omnia.* Ed. Carolus Gottlob Kühn. 20 vols. Leipzig, 1821-1833; repr. Hildesheim, 1965.

――. *De nominibus medicis.* Ed. Max Meyerhoff and Joseph Schacht. *Abhandlungen der preussischen Akademie der Wissenschaften,* philosophisch-historische Klasse 1931, no. 3.

――. *De placitis Hippocratis et Platonis libri novem.* Ed. Iwan Müller. Leipzig, 1874.

――. *De usu partium.* Ed. Georg Helmreich. 2 vols. Leipzig, 1907-1909.

――. *Galeni Scripta minora.* Ed. Ioannes Marquardt, Iwan Müller, and Georg Helmreich. Leipzig, 1884-1893.

――. "Galens Schrift über die Siebenmonatskinder." Ed. and trans. from Arabic by Richard Walzer. *Rivista di Studi Orientali* 15 (1935), 323-357, and 16 (1935/7), 227.

———. *On Medical Experience.* Ed. and trans. from Arabic by Richard Walzer. Oxford, 1944.

———. *On the Natural Faculties.* Ed. and trans. Arthur J. Brock. Loeb Classical Library. 1916.

Other English Translations of Galen

———. *On the Affected Parts.* Trans. Rudolph Siegel. Basel and New York, 1976.

———. *On Anatomical Procedures.* Trans. Charles Singer. London, 1956.

———. *On Anatomical Procedures: The Later Books.* Trans. W. L. H. Duckworth. Cambridge, 1962.

———. *On the Anatomy of the Uterus.* Trans. Charles M. Goss. *Anatomical Record* 144 (1962), 77–84.

———. *On Anatomy of the Veins and Arteries.* Trans. Charles M. Goss. *Anatomical Record* 141 (1961), 355–366.

———. *Hygiene.* Trans. Robert Montraville Green. Springfield, Ill., 1951.

———. *On Marasmus.* Trans. Theoharis C. Theoharides. *Journal of the History of Medicine* 4 (1971), 369–390.

———. *On the Usefulness of the Parts of the Body.* Trans. Margaret Tallmadge May. 2 vols. Ithaca, N.Y., 1968.

HIPPOCRATES. *Hippocrate.* Ed. Robert Joly et al. Collection des Universites de France (Bude, Paris). Vol. 6, 1967, 1972; vol. 11, 1970.

———. *Hippocrates.* Ed. and trans. William H. S. Jones and Edward T. Withington. Loeb Classical Library. London, 1923–1931.

———. *Hippocratis opera quae feruntur omnia.* Ed. Hugo Kulewein. Vols. 1, 2. Leipzig, 1894, 1902. No more published.

———. *Oeuvres complètes d'Hippocrate.* Ed. Emile Littré. 10 vols. Paris, 1839–1861.

———. *Opera.* Ed. I. L. Heiberg. Vol. 1. Leipzig, 1927. No more published.

———. *Schriften, Die Anfänge der Abendlandische Medizin.* Trans. Hans Diller. Hamburg, 1962.

———. *The Genuine Works of Hippocrates.* Trans. Francis Adams. London, 1849.

ORIBASIUS. *Oribasii Collectionum medicarum reliquiae.* Ed. Johannes Raeder. Leipzig, 1928–1933.

RUFUS OF EPHESUS. *Die Fragen des Aerztes an den Kranken.* Ed. Heinrich Gaertner. Berlin, 1962.

_____. *Oeuvres de Rufus d'Ephèse.* Ed. Charles Daremberg and Emile Ruelle. Paris, 1879; repr. Amsterdam, 1973.

II. Books by Modern Authors

ACKERKNECHT, ERWIN H. *Medicine at the Paris Hospital, 1794–1848.* Baltimore, 1961.
ALLBUTT, T. CLIFFORD. *Greek Medicine in Rome.* London, 1921; repr. New York, 1970.
BACON, FRANCIS. *The Advancement of Learning and the New Atlantis.* Ed. Arthur Johnston. Oxford, 1974.
_____. *The Works of Francis Bacon.* Ed. with a Life of the Author by Basil Montagu. 3 vols. Philadelphia, 1842.
BAILLOU, GUILLAUME DE. *De virginum et mulierum 'morbis liber.* Ed. M. Jacobus Thevart. Paris, 1643.
_____. *Epidemiarum et Ephemeridum libri duo.* Ed. M. Jacobus Thevart. Paris, 1640.
_____. *Opuscula Medica.* Ed. M. Jacobus Thevart. Paris, 1640.
BOERHAAVE, HERMANN. *Institutiones Medicae.* Leiden, 1713.
BOURGEY, LOUIS. *Observation et experience chez les médecins de la collection hippocratique.* Paris, 1953.
BOWERSOCK, GLEN W. *Greek Sophists in the Roman Empire.* Oxford, 1969.
BROCK, ARTHUR J., trans. *Greek Medicine: Being Extracts Illustrative of Medical Writers from Hippocrates to Galen.* London and New York, 1929.
CASTIGLIONI, ARTURO. *History of Medicine.* Trans. D. B. Krumbhaar. New York, 1947.
La collection hippocratique et son rôle dans l'histoire de la médecine; Colloque de Strasbourg 23–27 octobre 1972. Leiden, 1975.
Corpus Hippocraticum; Actes du colloque hippocratique de Mons 22–26 septembre 1975. Ed. Robert Joly. Mons, 1977.
DALY, LLOYD W. *Contributions to a History of Alphabetization in Antiquity and the Middle Ages.* Collection Latomus 90. Brussels, 1967.
DEICHGRÄBER, KARL. *Die Epidemien und das Corpus Hippocraticum. Abhandlungen der preussischen Akademie der Wissenschaften, philosophisch-historische Klasse,* no. 3. Berlin, 1933.
_____. *Die griechische Empirikerschule.* Berlin, 1930.
DEWHURST, KENNETH. *Dr. Thomas Sydenham, His Life and Original Writings.* Berkeley, 1966.
DILLER, HANS. *Die Ueberlieferung der hippokratischen Schrift Peri aeron hydaton topon.* Philologus Supplementband 23.3. Leipzig, 1932.

_____. *Wanderarzt und Aitiologe. Philologus Supplementband* 26.3. Leipzig, 1934.

EDELSTEIN, LUDWIG. *Ancient Medicine: Selected Papers.* Ed. Owsei Temkin and C. Lilian Temkin. Baltimore, 1967.

_____. *Peri Aeron und die Sammlung der hippokratischen Schriften.* Berlin, 1931.

FABRICIUS, CAJUS. *Galens Exzerpte aus älteren Pharmakologen.* Berlin, 1972.

FRASER, PETER M. *Ptolemaic Alexandria.* 2 vols. Oxford, 1972.

FREDRICH, CARL. *Hippokratische Untersuchungen. Philologische Untersuchungen* 15. Berlin, 1899.

GREEN, R. MONTRAVILLE. *Asclepiades, His Life and Writings.* New Haven, 1955.

GRENSEMANN, HERMANN. *Knidische Medizin* I. Berlin, 1975.

HARIG, GEORG. *Bestimmung der Intensität im medizinischen System Galens.* Berlin, 1974.

HARRIS, C. R. S. *The Heart and Vascular System in Ancient Greek Medicine.* Oxford, 1973.

HELMONT, JEAN BAPTISTE VAN. *Aufgang der Arztney-Kunst.* Sulzbach, 1683; repr. Munich, 1971.

_____. *Ortus Medicinae.* Amsterdam, 1648.

HERZOG, RUDOLPH. *Heilige Gesetze von Kos.* Berlin, 1928.

_____. *Koische Forschungen und Funde.* Leipzig, 1899.

JAEGER, WERNER W. *Diokles von Karystos.* Berlin, 1938.

JOLY, ROBERT. *Recherches sur le traité pseudo-hippocratique du régime.* Paris, 1960.

JOUANNA, JACQUES. *Hippocrate, pour une archéologie de l'école de Cnide.* Paris, 1974.

KOLLESCH, JUTTA. *Untersuchungen zu den pseudogalenischen Definitiones Medicae.* Berlin, 1973.

LE CLERC, DANIEL. *Histoire de la médecine.* The Hague, 1729.

LICHTENTHAELER, CHARLES. *Deux conférences.* Geneva, 1959.

_____. *La médecine hippocratique I.* Lausanne, 1948.

_____. *La médecine hippocratique II–V.* Boudry, 1957.

_____. *Quatrième série d'études hippocratiques.* Geneva and Paris, 1963.

LINDEBOOM, GERRITT A. *Bibliographia Boerhaaviana.* Leiden, 1959.

_____. *Herman Boerhaave, The Man and His Work.* London, 1968.

LITTRÉ, EMILE. *Médecine et médecins.* Paris, 1872.

MERCURIALIS, HIERONYMUS. *Censura operum Hippocratis.* Venice, 1585.

MEYER-STEINEG, THEODOR, AND SUDHOFF, KARL. *Geschichte der Medizin.* 4th ed. Jena, 1950.

NACHMANSON, ERNST. *Erotianstudien.* Uppsala, 1917.

NEUBURGER, MAX. *Geschichte der Medizin.* Stuttgart, 1906. Trans. Ernest Playfair as *History of Medicine.* London, 1910.
PAGEL, WALTER. *Paracelsus.* Boston and New York, 1958.
PARACELSUS. *Four Treatises of Theophrastus von Hohenheim, called Paracelsus.* Ed. Henry E. Sigerist. Baltimore, 1941.
_____. *Sämtliche Werke.* Ed. Karl Sudhoff. Berlin, 1931.
_____. *Theophrastus Paracelsus Werke.* Ed. Will E. Peuckert. Basel, 1965–1968.
POHLENZ, MAX. *Hippokrates und die Begründing der wissenschaftlichen Medizin.* Berlin, 1938.
SCHOENER, ERICH. *Das Viererschema in der antiken Humoralpathologie.* *Sudhoffs Archiv* Supplement 4. Wiesbaden, 1964.
SCHUBRING, KONRAD. "Bemerkungen zu der Galenausgabe von Karl Gottlob Kühn und zu ihrem Nachdruck: Bibliographische Hinweise zu Galen." In *Claudii Galeni opera omnia.* Repr. Hildersheim, 1965.
SEZGIN, FUAT. *Geschichte des arabischen Schrifttums.* Leiden, 1970.
SINGER, CHARLES. *A Short History of Medicine.* Oxford, 1928.
SPRENGEL, KURT. *Apologie des Hippokrates und seines Grundsätze.* Leipzig, 1789.
_____. *Versuch einer pragmatischen Geschichte der Arztneikunde.* Halle, 1792–1799.
STECKERL, FRITZ, ed and trans. *The Fragments of Praxagoras of Cos and his School.* Leiden, 1958.
SYDENHAM, THOMAS. *Opera omnia.* Ed. William A. Greenhill. London, 1884.
TEMKIN, OWSEI. *Galenism: Rise and Decline of a Medical Philosophy.* Ithaca, New York, 1973.
WALZER, RICHARD. *Galen on Jews and Christians.* Oxford, 1949.
WELLMANN, MAX. *Die Fragmente der sikelischen Aerzte, Akron, Philistion, und des Diokles von Karystos.* Berlin, 1901.
_____. *Hippokratesglossare. Quellen und Studien zur Geschichte der Naturwissenschaften und der Medizin 2.* Berlin, 1931.
_____. *Die pneumatische Schule bis auf Archigenes.* Philologische Untersuchungen 14. Berlin, 1895.

III. Articles

ACKERKNECHT, ERWIN H. "Aspects of the History of Therapeutics." *Bulletin of the History of Medicine* 36 (1962), 389–419.
ALEXANDERSON, BENGT. "Bemerkungen zu Galens Epidemienkommentaren." *Eranos* 65 (1967), 118–145.

BARDONG, KURT. "Beiträge zur Hippokrates- und Galenforschung." *Nachrichten von der Akademie der Wissenschaften in Göttingen,* philosophisch-historische Klasse, no. 7 (1942), 577–640.

BLASS, FRIEDRICH. "Die pseudhippokratische Schrift *Peri Physon* und der Anonymus Londinensis." *Hermes* 36 (1901), 405–410.

BOUSQUET, JEAN. "Inscriptions de Delphes." *Bulletin de Correspondance Hellenique* 80 (1956), 579–593.

BROECKER, L. O. "Die Methoden Galens in der literarischen Kritik." *Rheinisches Museum* 40 (1885), 415–438.

BRUNN, WALTER L. VON. "Betrachtungen über Hohenheims Kommentare zu den *Aphorismen* des Hippokrates." *Nova Acta Paracelsica* 3 (1946), 24–42.

DE LACY, PHILLIP. "Galen's Platonism." *American Journal of Philology* 93 (1972), 27–39.

DIELS, HERMANN. "Hippokratische Forschungen 1–4." *Hermes* 45 (1910), 125–150; 46 (1911), 261–285; 48 (1913), 378–407; 53 (1918), 57–87.

———. "Ueber die Excerpte von Menons Iatrika." *Hermes* 28 (1893), 407–434.

———. "Ueber einen neuen Versuch, die Echtheit einiger Hippokratischen Schriften nachzuweisen." *Sitzungsberichte der preussischen Akademie der Wissenschaften,* philosophisch-historische Klasse (1910), 1140–1155.

DILLER, HANS. "Der innere Zusammenhang der hippokratische Schrift *de victu.*" *Hermes* 87 (1959), 39–56.

———. "Eine stoisch-pneumatische Schrift im Corpus Hippocraticum." *Sudhoffs Archiv* 29 (1936), 178–195.

———. "Stand und Aufgaben der Hippokratesforschung." *Jahrbuch der Akademie der Wissenschaften und der Literatur* (Mainz, 1959), 271–287.

———. "Zur Hippokratesauffassung des Galenos." *Hermes* 68 (1933), 167–182.

DOBSON, JOHN F. "Erasistratus." *Proceedings of the Royal Society of Medicine,* Section of the History of Medicine 20 (1927), 21–28.

———. "Herophilus of Alexandria." *Proceedings of the Royal Society of Medicine,* Section of the History of Medicine 18 (1925), 19–32.

EDELSTEIN, LUDWIG. "The Genuine Works of Hippocrates." *Bulletin of the History of Medicine* 7 (1939), 236–248.

———. "Hippokrates" in *RE* Supplement vol. 6 (1935), 1290–1345.

FRASER, PETER M. "The Career of Erasistratus of Ceos." Istituto Lombardo di Scienze e Lettere di Milano, Classe di Lettere e Scienze Morale e Storiche, *Rendiconti* 103 (1969), 518–537.

FUCHS, ROBERT. "De Erasistrato capita selecta." *Hermes* 29 (1894), 171–203.

GRENSEMANN, HERMANN. "Polybus als Verfasser hippokratischen Schriften." *Abhandlungen der Akademie der Wissenschaften und der Literatur,* Mainz, geistes und sozialwissenschaftliche Klasse, no. 2 (1968), 1–18.

HEINIMANN, FELIX. "Diokles von Karystos und der prophylaktische Brief an König Antigonos." *Museum Helveticum* 12 (1955), 158–172.

ILBERG, JOHANNES. "Aus Galens Praxis." *Neue Jahrbücher für Klassische Altertum* 15 (1905), 276–312.

_____. "De Galeni vocum hippocraticorum glossario." In *Commentationes Ribbeck,* pp. 327–354. Leipzig, 1888.

_____. "Die Aerzteschule von Knidos." *Berichte über die Verhandlungen der sächsische Akademie der Wissenschaften zu Leipzig* 76, no 3. Leipzig, 1925.

_____. "Die Hippokratesausgaben des Artemidorus Kapiton und Dioskurides." *Rheinisches Museum* 45 (1890), 111–137.

_____. "Rufus von Ephesos." *Abhandlungen der sächsische Akademie der Wissenschaften,* philosophisch- historische Klasse 41.1. Leipzig, 1930.

_____. "Ueber die Schriftstellerei des Klaudios Galenos." *Rheinisches Museum* 44 (1889), 207–239; 47 (1892), 489–514; 51 (1896), 165–196: 52 (1897), 591–623.

JOLY, ROBERT. "La question hippocratique et le témoignage de Phèdre." *Revue des Etudes Grecques* 74 (1961), 69–92.

JONES, WILLIAM H. S. "Hippocrates and the Corpus Hippocraticum." *Proceedings of the British Academy* 31 (1948), 1–23.

JOUANNA, JACQUES. "Le médecin Polybe. Est-il l'auteur de plusieurs ouvrages de la collection hippocratique?" *Revue des Etudes Grecques* 82 (1969), 552–562.

KOLLESCH, JUTTA. "Galen und seine ärztlichen Kollegen." *Das Altertum* 11 (1965), 47–53.

KUDLIEN, FRIDOLF. "Herophilos und der Beginn der medizinischen Skepsis." *Gesnerus* 21 (1964), 1–13.

_____. "Posidonius und die Aerzteschule der Pneumatiker." *Hermes* 90 (1962), 419–429.

_____. "Probleme um Diokles von Karystos." *Sudhoffs Archiv* 47 (1963), 456–464.

_____. "Untersuchungen zu Aretaios von Kappadokien." *Abhandlungen der Akademie der Wissenschaften,* Mainz, geistes und sozialwissenschaftliche Klasse, no. 11 (1963), 1151–1230.

LLOYD, GEOFFREY E. R. "A Note on Erasistratus of Ceos." *Journal of Hellenic Studies* 95 (1975), 172–175.

LONIE, IAIN M. "The Cnidian Treatises of the Corpus Hippocraticum." *Classical Quarterly* 15 (1965), 1–30.

_____. "Cos versus Cnidus and the Historians: Part 1." *History of Science* 15 (1978), 42–75; "Part 2," pp. 77–92.

_____. "The Hippocratic Treatise *peri diaites oxeon*." *Sudhoffs Archiv* 49 (1965), 50–79.

_____. "Medical Theory in Heraclides Ponticus." *Mnemosyne* 18 (1965), 126–143.

LOVE, IRIS C. "Prelminary Report of Excavations at Cnidus 1970." *American Journal of Archaeology* 76 (1972), 61–76.

MEWALDT, JOHANNES. "Galenos über echte und unechte Hippokratika." *Hermes* 44 (1909), 111–134.

MÜLLER, IWAN VON. "Ueber die dem Galenos zugeschreibene Abhandlung *Peri tes aristes haireseos*." *Sitzungsberichte der Münchner Akademie der Wissenschaften*, philologisch- historische Klasse (1898), 53–162.

NUTTON, VIVIEN. "The Chronology of Galen's Early Career." *Classical Quarterly* 23 (1973), 158–171.

PFAFF, FRANZ. "Rufus aus Samaria, Hippokrateskommentator und Quelle Galens." *Hermes* 67 (1932), 356–359.

_____. "Die Ueberlieferung des Corpus Hippocraticum in der nachalexandrinischen Zeit." *Wiener Studien* 50 (1932), 67–82.

PHILIPPSON, ROBERT. "Verfasser und Abfassungszeit der sogennanten Hippokratesbriefe." *Rheinisches Museum* 77 (1928), 298–328.

POHLENZ, MAX. "Hippokrates." *Die Antike* 15 (1939), 1–18.

POMTOW, HEINRICH. "Delphische Neufunde III. Hippokrates und die Asklepiaden in Delphi." *Klio* 15 (1918), 303–338.

PREMUDA, LORIS. "Il magistero d' Ippocrate nell' interpretazione critica e nel pensiero filosophico di Galeno." *Annali dell' Università di Ferrara* n.s., sect. 1 (1954), 67–92.

SCARBOROUGH, JOHN. "Celsus on Human Vivisection at Ptolemaic Alexandria." *Clio Medica* 11 (1976), 25–38.

SCHÖNE, HERMANN. "Echte Hippokratesschriften." *Deutsche medizinische Wochenschrift* 36 (1910), 418–466.

SIGERIST, HENRY. "On Hippocrates." *Bulletin of the History of Medicine* 2 (1934), 190–214.

SMITH, WESLEY D. "Galen on Coans vs. Cnidians." *Bulletin of the History of Medicine* 47 (1973), 569–585.

STECKERL, FRITZ. "Plato, Hippocrates and the Menon Papyrus." *Classical Philology* 40 (1945), 166–180.

SUDHOFF, KARL. "Kos und Knidos." *Münchner Beiträge zur Geschichte der Naturwissenschaften und Medizin* IV/V (1927).

TEMKIN, OWSEI. "Celsus' 'On Medicine' and the Ancient Medical Sects." *Bulletin of the History of Medicine* 3 (1935), 249–264.

——. "Der systematische Zusammenhang im Corpus Hippocraticum."
Kyklos 1 (1928), 9–43.

——. "On Galen's Pneumatology." *Gesnerus* 8 (1951), 180–189.

WALSH, JOSEPH. "Galen's Discovery and Promulgation of the Function
of the Recurrent Laryngeal Nerve." *Annals of Medical History* 8
(1926), 176–184.

——. "Refutation of the Charges of Cowardice Made against Galen."
Annals of Medical History n.s., 3 (1931), 195–208.

WELLMANN, MAX. "Hippokrates des Herakleides Sohn." *Hermes* 64
(1929), 16–21.

——. "Hippokrates des Thessalos Sohn." *Hermes* 61 (1926), 329–334.

——. "Die Medizin bis in die zweite Hälfte des zweiten Jahrhunderts."
in Franz Susemihl, *Geschichte der griechischen Literatur in der Alexanderzeit* I, pp. 777–828. Leipzig, 1891.

——. "Die spätern Aerzte." In Franz Susemihl, *Geschichte der griechischen
Literatur in der Alexanderzeit* II, pp. 414–447. Leipzig, 1892.

——. "Das Hygieinon des Hippokrates." *Quellen und Studien zur Geschichte der Naturwissenschaften und der Medizin* 4.1 (1935), 1–5.

——. "Zur Geschichte der Medicin im Altertum." *Hermes* 47 (1912),
4–17.

WENKEBACH, ERNST. "Der Hippokratische Arzt als das Ideal Galens."
*Quellen und Studien zur Geschichte der Naturwissenschaften und der
Medizin* 3.4 (1932–1933), 155–175.

INDEX

Ackerknecht, Erwin H., 31-33, 96n
Adams, Francis, 32
Aelius Aristides, 63
Aephicianus, 64, 67, 70, 91, 118,
    147-148, 179, 226, 245
Alexandria (Alexandrians), 34, 59, 62,
    72, 190, 191, 199, 202, 206n, 212,
    223, 230, 232, 235, 237, 245
Allbutt, Sir T. Clifford, 30n, 198n
Anatomy, 65-69, 78-79, 83-85,
    149-150, 171, 189-90, 195, 208,
    213, 246; vivisection, 190, 194,
    208
Anaxagoras, 45, 49
Andreas, 206n, 211, 233
Anonymus Londinensis. See Menon
Antigenes, 78, 82
Apollonius, father and son, Empirics,
    211, 212
Apollonius, student of Hippocrates,
    95, 115, 136-138, 180-181, 196
Apollonius of Citium, 128n, 212-215,
    222, 245
Archigenes, 80, 101, 104, 118,
    230-234
Aretaeus of Cappadocia, 25, 230,
    243-245
Argentier, 18
Aristarchus, 161, 235
Ariston, 116-117, 197, 239
Aristophanes of Byzantium, 161, 202

Aristotle, 24, 27, 35, 55-56, 75, 84-87,
    92-93, 100, 105, 109n, 139n, 168,
    181, 186, 189, 194, 211, 220-221,
    223, 231-232
Artemidorus Capiton, 122, 127, 133,
    146, 148, 153, 158, 170n, 173,
    179, 220, 226, 234-240
Asclepiadae (also Asclepiads), 45, 128,
    216-218, 222-223, 237
Asclepiades of Bythinia, 75-76, 80,
    103, 105, 146-148, 175, 178, 182,
    190, 193, 198, 205, 223-229, 232,
    240n
Asclepius, 62-63, 211
Athenaeus of Attaleia, 80, 87, 100,
    117-118, 139n, 231-234
Athenaeus of Naucratis, 98, 225n

Bacchius, 130, 141, 146, 150, 161, 191,
    194n, 202-204, 210, 212, 214,
    234, 243
Bacon, Francis, 18, 19, 20, 24, 50
Baillou, Guillaume de, 18-19, 25, 26n
Bardong, Kurt, 61n, 83n, 95, 110n,
    119n, 189n, 236
Bayle, Antoine Laurent, 18
Blass, Friedrich, 37, 40, 58
Bloodletting (phlebotomy,
    venesection), 21, 32, 63n, 73,
    79-82, 143
Boerhaave, Hermann, 23-26, 243

257

Library of Congress Cataloging in Publication Data
(For library cataloging purposes only)

SMITH, WESLEY D 1930–
    The Hippocratic tradition.

    (Cornell publications in the history of science)
    Bibliography: p.
    Includes index.
    1. Hippocrates. 2. Galenus. 3. Medicine, Greek
and Roman. I. Title. II. Series: Cornell University.
Cornell publications in the history of science.
R126.H8S57   610'.938   78-20977
ISBN 0-8014-1209-9